NANOTECHNOLOGY SCIENCE AND TECHNOLOGY

ZnO NANOSTRUCTURES

ADVANCES IN RESEARCH AND APPLICATIONS

NANOTECHNOLOGY SCIENCE AND TECHNOLOGY

Additional books and e-books in this series can be found on Nova's website under the Series tab.

NANOTECHNOLOGY SCIENCE AND TECHNOLOGY

ZnO NANOSTRUCTURES

ADVANCES IN RESEARCH AND APPLICATIONS

DANA CRAWFORD
EDITOR

nova science publishers
New York

NOTICE TO THE READER

The Publisher has taken reasonable care in the preparation of this book, but makes no expressed or implied warranty of any kind and assumes no responsibility for any errors or omissions. No liability is assumed for incidental or consequential damages in connection with or arising out of information contained in this book. The Publisher shall not be liable for any special, consequential, or exemplary damages resulting, in whole or in part, from the readers' use of, or reliance upon, this material. Any parts of this book based on government reports are so indicated and copyright is claimed for those parts to the extent applicable to compilations of such works.

Independent verification should be sought for any data, advice or recommendations contained in this book. In addition, no responsibility is assumed by the Publisher for any injury and/or damage to persons or property arising from any methods, products, instructions, ideas or otherwise contained in this publication.

This publication is designed to provide accurate and authoritative information with regard to the subject matter covered herein. It is sold with the clear understanding that the Publisher is not engaged in rendering legal or any other professional services. If legal or any other expert assistance is required, the services of a competent person should be sought. FROM A DECLARATION OF PARTICIPANTS JOINTLY ADOPTED BY A COMMITTEE OF THE AMERICAN BAR ASSOCIATION AND A COMMITTEE OF PUBLISHERS.

Additional color graphics may be available in the e-book version of this book.

Library of Congress Cataloging-in-Publication Data

ISBN: 978-1-53616-773-3

Published by Nova Science Publishers, Inc. † New York

CONTENTS

PREFACE

Different ZnO nanostructures are attributed to different electronic properties which are favourable in the application of photocatalysis, sensing and energy harvesting. The non toxic behaviour of ZnO nanostructures lead to the use of this material in environmental sensing and environmental remediation.

In ZnO Nanostructures: Advances in Research and Applications, a large-scale synthesis of undoped low-dimensional semiconductor metal oxide nanostructures is performed by simple wet-chemical method using reducing agents at low temperature. The detailed structural, compositional, and optical characterizations of the ZnO nanoparticles were evaluated by powder X-ray diffraction pattern, Fourier-transform infra-red spectroscopy, X-ray photoelectron spectroscopy, electron dispersion spectroscopy and UV-vis. Spectroscopy.

Following this, the authors describe the structure electrochemical sensors, listing scientific papers focusing on the detection of different pathogens and analytes, as well as reporting and comparing the performance of the sensors prepared by various groups around the world.

The concluding chapter deals with synthesis protocols of ZnO heterostructures along with their role in optoelectronic applications. Their thermodynamic stability and correlation among morphology, defects, and heterostructure with luminating and catalytic properties is also described.

Chapter 1 - ZnO nanostructures have been the most demanding material in last few decades due to the exhibition of enormous applications of this material in various fields. These nanostructures can be controlled dimensionally through variations in hydrothermal process. Different ZnO nanostructures are attributed to different electronic properties which is favourable in the application of photo catalysis, sensing, energy harvesting etc. The non toxic behaviour of ZnO nanostructures lead to the use of this material as environmental sensing and environmental remediation. One dimensional ZnO nanostructure shows very good hydrophobic and antibacterial property which has many applications in self-cleaning surface design industries.

Chapter 2 - In this approach, a large-scale synthesis of undoped low-dimensional semiconductor metal oxide nanostructures (ZnO nanoparticles, NPs) by simple wet-chemical method was performed using reducing agents at low temperature. The NPs were characterized in terms of their morphological, structural, and optical properties, and efficiently applied for the metal ions uptake. The detailed structural, compositional, and optical characterizations of the NPs were evaluated by powder X-ray diffraction pattern (XRD), Fourier-transform infra-red spectroscopy (FTIR), X-ray photoelectron spectroscopy (XPS), Electron dispersion spectroscopy (EDS), and UV-vis. spectroscopy, respectively which confirmed that the obtained NPs are well-crystalline undoped ZnO and possessed good optical properties. The ZnO NSs morphology was investigated by FESEM, which confirmed that the calcined materials were spherical shape in nano-lavel and growth in huge-quantity. The analytical efficiency of newly synthesized ZnO NPs was also investigated for a selective separation of trivalent iron [Fe(III)] prior to its determination by inductively coupled plasma-optical emission spectrometry (ICP-OES). The selectivity of ZnO NPs towards different metal ions, including Cd(II), Co(II), Cr(III), Cu(II), Fe(III), Ni(II), Zn(II), and Zr(IV), was studied. Data obtained from the selectivity study suggested that that ZnO NPs phase was the most selective towards Fe(III). The static uptake capacity of Fe(III) was found to be ~79.80 mgg^{-1}. Moreover, adsorption isotherm data also provided that the adsorption process was mainly monolayer on a homogeneous adsorbent surface.

Chapter 3 - Pathogens, infectious microbes that spread diseases, are one of the leading causes of death among youngs, causing several million deaths per year. Pathogens detection in human environment (especially indoor) is therefore crucial to reduce the risk and save human lives. Furthermore, small, cheap and portable devices, which can be used by personnel with little or no training, are crucial for early diagnosis in areas without hospital or points-of-care. Electrochemical sensors are ideal candidates for this role, and ZnO nanostructures are proving very useful for optimizing their performance. In this chapter the authors will briefly describe the structure of this type of biosensors, and then show the different types of ZnO nanostructures that are used in this field, grown through different methods. The authors will list scientific papers focusing on the detection of different pathogens and analytes, reporting and comparing the performance of the sensors prepared by various groups around the world. The use of nanoparticles, nanorods, nanowires, nanotubes, nanofibres, nanotetrapods and other morphologies at the nanometer level will be described. The electrochemical biosensors discussed in this chapter detect the concentration of glucose inside cells, presence of hepatitis B virus, E. coli, H1N1 swine influenza virus, cholesterol, Legionella pneumophila, uric acid, C-reactive proteins, N. meningitidis, different Leptospira species, urea, avian influence, and others. The authors think that this overview of electrochemical biosensors using ZnO nanostructures in one of their components can be useful for approaching this field.

Chapter 4 - ZnO, a well-abundant material, has a wide range of potentiality in various optoelectronic devices such as light emitting diode, solar cell, transducer, waveguide, sensor etc., attributed to its environmental biocompatibility, chemical stability, suitable band gap (~ 3.37 eV), electron mobility etc. Other important properties which make it attractive in diverse field are piezoelectricity, excitonic (~60 meV), photostability, electron donating and accepting property, photocatalytic etc. In general, all optoelectronic applications of ZnO primarily depend on band to band near edge and defect mediated electronic transitions. Defects, classified as intrinsic defects and extrinsic defects, can be varied easily in ZnO by synthesis conditions; hence by controlling defects, ZnO can be tailor made for various optoelectronic applications. Intrinsic defects get formed from either interstitial or deficiency of Zn and or O, while extrinsic defects are related with doping with other elements *viz.* P, Mn, Ni, Al, S, N etc. On the basis of the position of defect levels with respect to conduction and valence band edges, they are classified into deep level defects and shallow defects. In general, shallow defects contribute to electrical conductivity, while deep level defects significantly effects on optical transition giving luminating property and photocatalytic activity of ZnO. In recent time, it has been emphasized that optoelectronic properties of ZnO is modified widely in its nanostructured form and much efforts have been given to understand the optoelectronic behaviour of ZnO, followed by device fabrication. Various techniques like hydrothermal, solvothermal, sol-gel, chemical bath deposition, physical vapour deposition, atomic layer deposition, pulsed laser deposition have been adopted to generate different nanostructure of ZnO such as nanorods, nanoplate, nano-flower, hedge – hog etc. In general, all these synthesis processes introduce different kinds of defect, morphology etc. that impact differently on optoelectronic properties of ZnO nanomaterials, particularly in fabricating light emitting diode. Thus, to gain insight into defect mediated application aspects of ZnO nanostructured materials, their correlation with the defect structure is highly essential and it is being elaborated within this chapter. ZnO nanostructures are also being used in degradation of toxic dyes by photocatalytic process, ascribed to suitable energy position of conduction and valence band edges to generate reactive oxygen species. Here also defects play an important role to inhibit electron – hole recombination, which successively increases catalytic efficiency. In this context, it may be stated that stability of the defects as determined by thermodynamics is important. Nowadays, efforts are given to tune optoelectronic properties of ZnO by forming its heterostructure, where photo-excited electron transfer across the junction plays crucial role in modifying the optoelectronic properties of ZnO in light emitting devices and catalytic applications. This present chapter also deals with synthesis protocols of ZnO heterostructures along with its role in these optoelectronic applications. In this context, it may be stated primarily that charge separation across the heterostructure which causes less recombination of photogenerated electron and hole tunes plays important role to tune emission and catalytic efficiency. In brief, present chapter highlights thermodynamic stability, and correlation among morphology, defects, heterostructure with luminating and catalytic properties.

In: ZnO Nanostructures
Editor: Dana Crawford
ISBN: 978-1-53616-773-3

Chapter 1

ZnO Nanostructures for Environmental Applications

***Arnab Kumar Sarkar*[1], *Rashi Borgohain*[2]**
***Bikash Agarwal*[3] *and Sunandan Baruah*[4,*]**
[1]Department of Electronics and Communication Technology, Gauhati University, Guwahati, Assam, India
[2]Department of Electronics and Telecommunication Engineering, Assam Engineering College, Guwahati, Assam, India
[3]Department of Electronics and Communication Engineering, Assam Don Bosco University, Guwahati, Assam, India
[4]Centre of Excellence in Nanotechnology, Assam Don Bosco University, Guwahati, Assam, India

Abstract

ZnO nanostructures have been the most demanding material in last few decades due to the exhibition of enormous applications of this material

---

[*] Corresponding Author's Email: sunandan.baruah@dbuniversity.ac.in.

in various fields. These nanostructures can be controlled dimensionally through variations in hydrothermal process. Different ZnO nanostructures are attributed to different electronic properties which is favourable in the application of photo catalysis, sensing, energy harvesting etc. The non toxic behaviour of ZnO nanostructures lead to the use of this material as environmental sensing and environmental remediation. One dimensional ZnO nanostructure shows very good hydrophobic and antibacterial property which has many applications in self-cleaning surface design industries.

Keywords: zinc oxide, hydrothermal growth, environmental sensor, remediation, photocatalysis, dye sensitized solar cell, piezotronic

1. INTRODUCTION

ZnO nanomaterial has been a subject of focused research nowadays due to its unique properties that are significantly different from those of bulk counterpart. It has various morphologies which can be synthesized by different simple and low cost growth methods.

1.1. Structure

ZnO has wurtzite crystal structure as shown in Figure 1.1. The lattice parameters of hexagonal zinc oxide are a = 0.3296 nm and c = 0.52065 nm [1]. In the crystal, zinc and oxygen planes along the c-axis are tetrahedrally coordinated (Figure 1.1.). The non-central symmetric structure of ZnO is the result of the tetrahedral coordination and consequently results in piezoelectric and pyroelectric property. Having polar surfaces are another important characteristic of ZnO. The basal plane (0001) of ZnO which is a polar surface, alternately terminated by positively charged zinc ions and negatively charged oxygen ions resulting in a normal dipole moment and spontaneous polarization along the c-axis. The other surfaces of ZnO are non-polar which is shown in Figure 1.2.

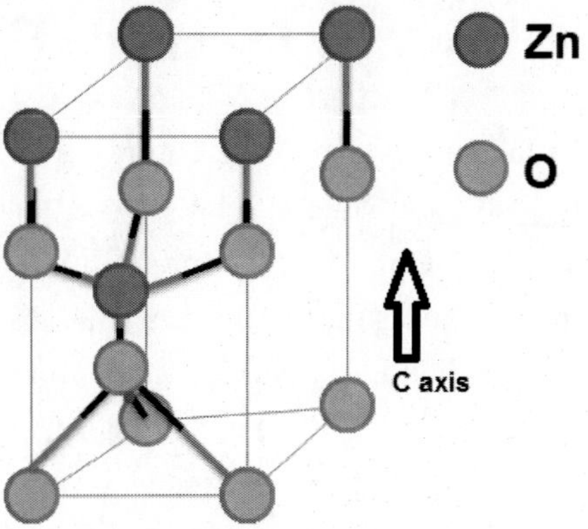

Figure 1.1. Wurtzite structure of ZnO .

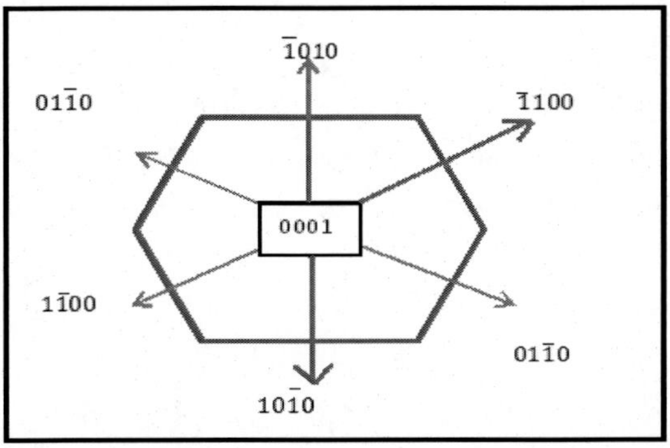

Figure 1.2. Different facets of Wurzite structure of ZnO.

1.2. Properties

The direct band gap and exciton binding energy of ZnO are 3.37 eV and 60 meV respectively at room temperature [2].Among the nanostructured sensing materials, ZnO has been usedas a highly useful sensing material. Because, ZnO is a multifunctional material and has many advantages like high electrochemical stability, non-toxicity, suitable for doping, biocompatible,high electrochemical coupling coefficient, high photo-stability and also economical over other nanomaterials [3].

1.3. Synthesis

Synthesis of nanomaterials with exact control over size, shape, and crystalline structure has become very important for different nanotechnology applications. Top down technique involves the division of a massive solid into smaller and smaller portions, successively reaching to nanometer size. On the other hand, in bottom up technique nanocomponents are fabricated using chemicals using either chemical or physical deposition processes. Bottom up technique is more chosen because it is cheap and suitable for large scale production compared to top down technique. There are many bottom up methods of synthesizing nanomaterials, such as hydrothermal, [4, 5] combustion synthesis [6], gas-phase methods [7, 8], sol-gel processing [9], etc.

Compared to other methodologies, hydrothermal synthesis is eco-friendly and cheap [10]. It isa low temperature process without the requirement of any sophisticated setup like high temperature and high vacuum reaction chambers.It is normally carried out in an autoclave with the reaction in aqueous solution [11]. This method has the advantage of controlling the grain size, particle morphology, crystalline structure and surface chemistry of the nanoparticles [12]. Therefore, hydrothermal synthesis is widely used for the synthesis of metal oxide nanoparticles. This process can be carried out using both convection and microwave heating.

1.3.1. Hydrothermal Synthesis with Microwave Irradiation

Microwave synthesis is relatively new and an interesting technique for the synthesis of nanomaterials [13]. In this process nanomaterials are synthesized in remarkably short durations under microwave irradiation [14, 15]. Utilizing microwave energy for the thermal treatment generally leads to homogeneous reaction condition producing a very fine particle in the nanocrystalline regime.

1.3.2. Synthesis of ZnO Nanoparticles

For the synthesis of ZnO nanoparticles, 4 mM zinc acetate dihydrate [$Zn(CH_3COO)_2.2H_2O$, Merck, 99 per cent purity] solution is to be prepared in 20 ml of ethanol [C_2H_5OH, Merck, 99 per cent purity] with vigorous stirring at 50°C. The solution is then diluted with another 20 ml of fresh ethanol and cooled in the ambient air, following which 20 ml of 4 mM sodium hydroxide in ethanol is added drop-wise to the solution under continuous stirring. The mixture is then kept in a temperature controlled water bath at 60°C for 2 hours and after that cooled to room temperature [1, 3, 6]. The reactions leading to the formation of the ZnO nanoparticles can be explained with the following equations

$$Zn(CH_3COO)_2.2H_2O + 2NaOH \rightarrow Zn(OH)_2 + 2CH_3COONa + 2H_2O \quad (1.1)$$

$$Zn(OH)_2 + C_2H_5OH \rightarrow Zn^{2+} + 2OH^- + 2C_2H_5OH \quad (1.2)$$

$$Zn(OH)_2 + C_2H_5OH \rightarrow Zn(OH)_4^{2-} + 2C_2H_5{}^+ \quad (1.3)$$

$$Zn(OH)_4^{2-} \rightarrow ZnO + H_2O + 2OH^- \quad (1.4)$$

1.3.3. Growth of ZnO Nanorods

The seeding of the particles can be carried out on different substrates by dipping them in the synthesized dispersion of ZnO nanoparticles for 15 minutes. Three such dipping are required and after each dipping the substrate is washed with deionized water to remove the loosely attached particles and then it is heated at 150°C for 15 minutes. Preheating of the substrate is done at 100°C and post annealing at 150°C.The growth of ZnO nanorods on the seeded substrate is carried out in a hot air oven maintained at 90°C. The substrate is placed inverted in a petri dish containing equimolar solution of zinc nitrate hexahydrate [($Zn(NO_3)_2.6H_2O$), Merck, 99 per cent purity] and hexamethylenetetramine [($C_6H_{12}N_4$), Merck, 99 per cent purity] and then kept in the oven for 15 hours. After every 5 hours the solution mixture is to be changed and at the completion the substrate is washed thoroughly with deionized water [1].

2. Diagnosis of Environmental Health

A safe, healthy and supportive environment is necessary to promote health for all human beings.Environmental health refers toprevent or control disease, injury, and disability related to the interactions between people and their environment. The greatest hindrance to the environmental health is pollution. It has become a serious threat to our existence.It is increasing exponentially causing serious and irreparable damage to the air we breathe, the water we drink and the soil which produces our crops, vegetables and fruits, adversely affecting our health.

Air pollution is caused by the toxic smoke emitted by cars, buses, trucks, trains, and factories, namely sulphur dioxide, carbon monoxide and nitrogen dioxide, etc. Evidence of increasing pollution is manifested by the increased rate of patients with lung cancer, asthma, allergies, and various breathing problems along with severe and irreparable damage to flora and fauna [16]. Even the most natural phenomenon of migratory birds has been hampered, with severe pollution preventing them from reaching their seasonal metropolitan destinations of centuries [16].

Heavy metals like lead, mercury, cadmium, arsenic, chromium, zinc, nickel and copper released from metal casting and refining industries, scrap metal, plastic and rubber industries, various consumer products and from burning of waste, are deposited onto the soil, vegetation and water, poisoning humans through inhalation, ingestion and skin absorption. Agricultural crops are injured on exposureto high concentrations of various pollutants, resulting in visible markings on the foliage, reduced growth and premature death of the plant [16].

The availability of clean and safe drinking water is of major concern throughout the world nowadays. There are enormous invisible microorganisms present in every drop of water that swim and multiply by billions.To prevent and control waterborne diseases, drinking water must be free from pathogens. The most common bacterial diseases transmitted through water are Acute diarrhea and gastroenteritis, Pneumonia, Cholera, Typhoid fever, Bacillary dysentery, etc. [17].

Therefore, for the diagnosis of environmental health early detection and monitoring of these poisonous and hazardous chemicals are utmost necessary.

2.1. Detection of Polluting Gases

One dimensional nanostructured semiconductor based gas sensors can be used for detecting the polluting gases as they have many advantages like robustness, lightweight, durable, highly sensitive and cheap [18]. They are being used widely to measure and monitor trace amounts of ecologically harmful gases such as carbon monoxide, nitrogen dioxide, etc. [19].

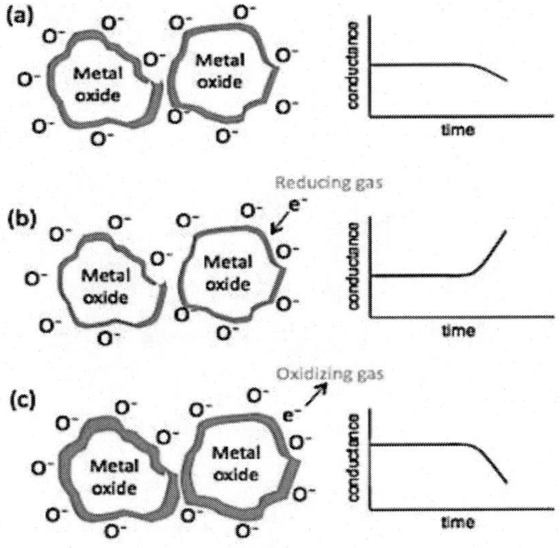

Figure 2.1.Conductance curve of metal oxide semiconductor in presence of (a) ambient air, (b) reducing gas, (c) oxidising gas.

2.1.1. Operating Principle of Metal Oxide Semiconductor as Gas Sensors

When semiconductors are used as gas sensors, there is a change in their conductivity with the variation in the concentration of the analyte gas. Nanostructures allow higher adsorption of gas molecules onto its surface due to their large surface area to volume ratio. One-dimensional metal oxide nanostructures like nanowires, nanorods, etc. provides more sensitivity and stability compared to other nanostructures[20]. Adsorption of ions of the analyte species over the surface of the sensing material is the basic phenomenon behind the sensing process.

As shown in Figure 2.1. (a), in ambient air, oxygen ions are adsorbed on the surface of the sensor and scavenge electrons in the process from the surface of the semiconducting sensing element. As a result of this electron extraction, the conductance of the semiconductor decreases which is expressed by equation (2.1)

$$O_2 + e^- \rightarrow O_2^- \tag{2.1}$$

where, O_2 is an oxygen molecule, e^- is a free electron, and O_2^- is an adsorbed oxygen molecule on the surface of the semiconductor.

Now, if this semiconducting material is exposed to a reducing gas, then the reducing gas molecules react with the adsorbed oxygen ions by releasing electrons to the semiconductor and thereby conductance of the semiconductor increases which is shown in Figure 2.1. (b) and expressed by equation (2.2).

$$R(g) + O^- \rightarrow RO(g) + e^- \tag{2.2}$$

where R is the reducing gas.

In presence of an oxidising gas, more oxygen ions are adsorbed onto the surface of the semiconductor by extracting electrons from its surface and as a result of which conductance further decreases which is shown in Figure 2.1. (c) and expressed by equation (2.3).

$$XO_2(g) + e^- \rightarrow XO + O^- \tag{2.3}$$

where XO_2 is an oxidising gas.

2.1.2.ZnO for Gas Sensing

Among the semiconducting materials, ZnO has attracted increasing attention and has been proven to be a highly useful sensing material for detecting both oxidizing and reducing gases. Hitherto, low dimensional ZnO nanostructures with different morphologies including nanobelts, nanotubes, nanorods, nanowires, nanofibers, nanodisks, nanospindles, and nano-needles have been successfully developed, and many exhibit interesting gas sensing performances towards H_2, CO, NO_2 and some volatile organic compounds [21].

Bai *et al.* [22] synthesized ZnO nanostructures with flower like morphology for sensing gases like NO_2, CO, CH_4 emitted by vehicle exhaust. During the synthesis process, zinc nitrate hexahydrate and sodium hydroxide were used as the raw materials. The authors observed that the morphology and aspect ratio of the nanostructures vary subject to annealing temperature. The best response was obtained at operating temperature of 150°C and annealing temperature of 600°C among the calcination temperature of 400°C, 600°C and 800°C.

Pawar *et al.* [23] synthesized ZnO nanorods using zinc acetate, hexamethylenetetramine and some surfactants such as polyethylenimine (PEI), polyacrylic acid (PAA), diaminopropane(DAP) and DAP+PAA. These surfactants play an important role in fine tuning the surface morphology of ZnO nanostructures. They found the aspect ratio of ZnO nanostructures increased a lot because of the addition of these surfactants. They also compared the response of these sensors to different gases but the response for acetone was significant, which was 84 at 325°C. The authors then compared the sensitivity of the different nanostructures to 2000 ppm acetone and found that the structure with DAP surfactant showed the highest response of 92 at 300°C.

It is well known from present semiconductor technologies that the incorporation of impurities or defects into the semiconductor lattices is the primary means of controlling electrical conductivity, optical, luminescent, magnetic, and other physical properties [24]. With the help of doping we can change the shape and size of nanostructures. It is usually done with noble metals, rare earth oxides and transition metals. Co-doping of transition metal ions such as Ni, Co, Mn are also established.

Shouli *et al.* [25] reported the synthesis of pencil-like, needle-like and flower-like morphologies of ZnO nanorod by hydrothermal process using CTAB, SDS and PEG respectively. In CTAB assisted process, using zinc nitrate hexahydrate ($Zn(NO_3)_2.6H_2O$) and sodium hydroxide (NaOH) uniform pencil-like nanorods were synthesized with diameter in the range of 80-100nm and length 3.2μm with an aspect ratio of 40:1. The same raw materials along with SDS were used to synthesize needle like ZnO nanorods with an aspect ratio of 50:1. Flower like ZnO nanorods were synthesized by using $Zn(AC)_2.2H_2O$ and $NH_3.H_2O$ with PEG assistance. The authors found that crystal morphology of the as synthesized ZnO depend on the synthesis approach as well as the surfactant used. They also observed the response of various morphologies to oxidising and reducing gases and found that response of flower- like ZnO nanorods to 40 ppm CO was higher than pencil-like nanorod and needle-like nanorod, and the response of pencil-like nanorod to 40 ppm NO_2 was higher than needle-like and flower-like nanorod. However, in both the cases the response of pencil-like nanorods and needle-like nanorods were closer to each other and far away from flower like nanorods. The response could be enhanced by maintaining the calcination temperatures and time and also by doping with Cd. After doping with Cd, the response of needle like ZnO nanorod to 40 ppm CO increased from 48.7 to 167.2 at 400°C and that of pencil like nanorod increased from 52 to 470 at 350°C.

Nowadays, core–shell structures have attracted considerable attention because of their ability to detect trace gases. These types of hetero-structures enhance the sensing performance such as gas response and selectivity. Moreover, single semiconductor oxides usually show low sensitivity, selectivity and reliability and therefore, composite nanostructures like ZnO/TiO_2, In_2O/SnO_2, ZnO/SnO_2, ZnO/ZnS, etc. have been introduced to enhance gas sensing performance [26].

Lu *et al.* [27] synthesized ZnO/SnO_2 heterostructure for NO_2 gas sensing. They synthesized ZnO nanorods using zinc nitrate hexahydrate and hexamethylenetetramine first and then uniformly distributed SnO_2 on the surface of ZnO nanorods by hydrolysing $SnCl_2$. This composite structure showed good response in presence of UV light. They studied the response by giving different molar ratios of ZnO to SnO_2 termed as ZS0, ZS1, ZS2,

ZS3, ZS4 and ZS100 where ZS0 is 10:1, ZS3 is 1:1 and ZS100 is 1:10 and found that 1:1 molar ratio gave the best response of 1266 to 500 ppb of NO_2 at room temperature. They also investigated the selectivity of the sensor and found that sensor showed higher response to NO_2 than the other gases in presence of UV light.

Hwang *et al.* [28] synthesized ZnO –SnO_2 core shell nanowire and got good selectivity and sensitivity to 10 ppm of NO_2 at 200°C. They also found that the hetero-structure responded better to NO_2 than ZnO nanorods at 200°C

Park *et al.* [29] synthesized Zn_2SnO_4/ZnO core/shell nanorods sensors and obtained a response of 173-498% to 5 ppm NO_2 at 300°C.

Borgohain *et al.* [30] synthesized ZnO/ZnS core/shell nanostructure for room temperature detection of NO_2. In Fig. 2.2, it is seen that sensor response increases from 1800 per cent for 100 ppb of NO_2 to 3000 per cent for 2400 ppb of NO_2 at room temperature. The minimum concentration the sensor can respond to was reported to be 100 ppb at room temperature. In figure 2.3., the reproducibility and repeatability of the sensor in presence of 488 ppb of gas were observed and exhibited consistency upon exposure and removal of NO_2.

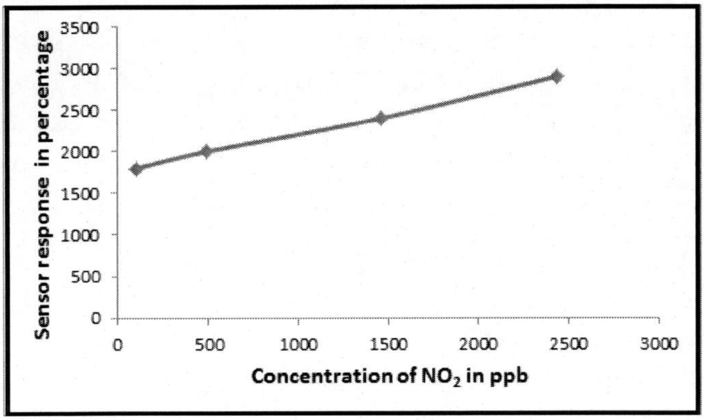

Figure 2.2. Response of the sensor to different concentration of NO_2 gas [30].

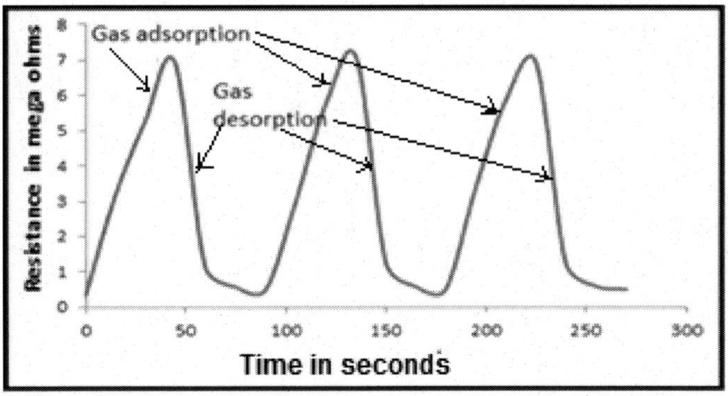

Figure 2.3. Repeatability of the sensor in presence of 488 ppb NO_2 [30].

2.2. Detection of Harmful Micro-Organisms in Water

Nanostructures can be used for detection of microorganisms present in water. The available methods for rapid identification of microorganisms present in water are colorimetric and fluorimetric methods, Genetic methods, DNA sequencing, Mass spectrometry, Microarrays, etc. [31]. These methods require sophisticated instruments which are very expensive and can be used only in controlled environment. Moreover they take considerable time for measurement [32]. In contrast to the expensive instrumentation, now-a-days, semiconducting nanostructures have been used as biosensors. The nanostructure-based devices can detect microorganisms with improved sensitivity, selectivity, and multiplexing capacity [33].

2.2.1. ZnO as Bacteria Sensor

Ivanova *et al.* [34] fabricated a thin film of ZnO and showed good interaction between Gram-positive bacteria of genus Bacillus and nanostructured thin film of ZnO. Tak *et al.*[35] detected single stranded DNA corresponding to *N. meningitides* using flower-like ZnO nanostructures.

Tereshchenko *et al.* [36] determined Grapevine virus A-type (GVA) proteins using a novel sensitive optical biosensor made up of ZnO thin films.

Borgohain *et al.* [37] fabricated sensors using ZnO nanorods for detection of two bacterial pathogens namely Escherichia Coli which is a gram negative bacteria and Streptococcus pneumonia which is a gram positive bacteria present in water at different concentrations. Figure 2.4. (a) shows TEM image of the ZnO nanoparticles, with individual particle sizes are about 8 nm on an average. The size and crystalline structure of ZnO nanoparticles can be obtained from the TEM image. The aspect ratio of the ZnO nanorods are about 100 with a length of 2μm and diameter of 20 nm approximately as shown in figure 2.4.(b). Figure 2.4. (c) and 2.4.(d) show the TEM image and selected area diffraction (SAED) pattern of the nanorods. The SAED pattern confirms the single crystalline structure of the ZnO nanorods.

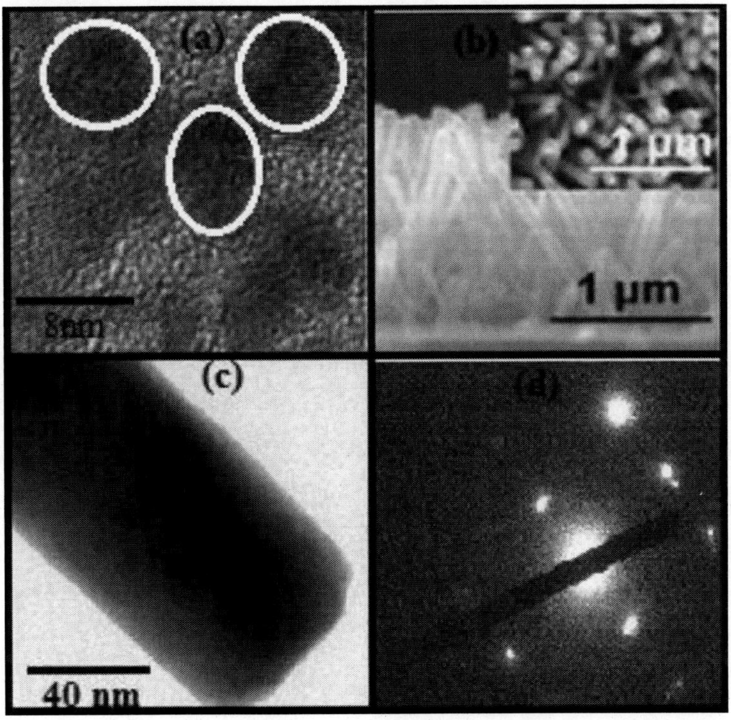

Figure 2.4. (a) TEM image of ZnO nanoparticles, (b) SEM image of ZnO nanorod (c) TEM image of ZnO nanorod, (d) SAED pattern of ZnO nanorod [37].

The concentrations of the pathogens used were 1.83X10^9 cells/ml and 2.0859X10^9 cells/ml for *E.coli* and *S. pneumonia* respectively. Different samples with percentage of bacteria in water 1.01, 1.12, 1.44, 1.69, 2.04, 11.1, 25, 50, 100 for both the pathogens were prepared for testing.

The results obtained from the sensor system are shown in figure 2.5. and 2.6. In figure 2.5., it is seen that with the increase in concentration of *E. coli* bacteria, the output voltage of the sensor system was found to be decreasing. This is because the bacteria cells are good conductor of current and that is why with the increase in concentration of bacteria the resistance of the sensor decreases and therefore voltage across the senor also decreases. In figure 2.6., it is seen that for *S. pneumonia* also sensor voltage was decreasing with the increase in concentration of the bacteria, but the rate of change was slower in Gram positive bacteria than gram negative bacteria.

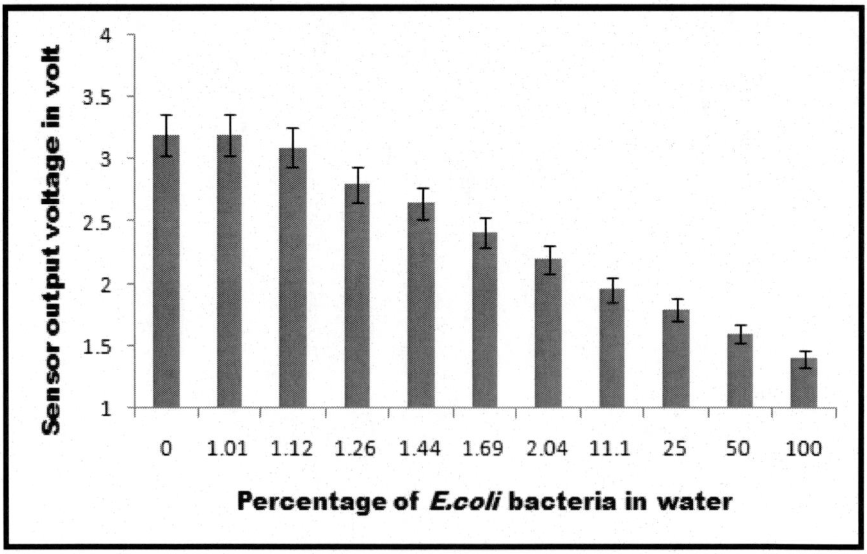

Figure 2.5. Output voltage of ZnO based sensor for *E. Coli* bacteria [37].

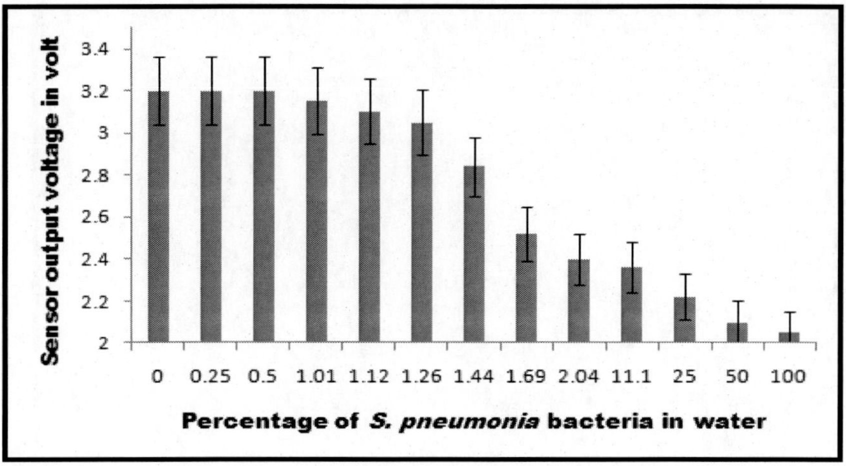

Figure 2.6. Output voltage of ZnO based sensor for *S. pneumonia* bacteria [37].

From figure 2.5. it is seen that, a little change was observed in the sensor output for the sample with 1.12 per cent bacteria, so this was the limit of detection of the sensor for *E.coli* bacteria. Similarly, from figure 2.6. we can see that the limit of detection of the sensor for *S. pneumonia* is 1.01 per cent.

Figure 2.7. and figure 2.8. shows the reproducibility and repeatability of the sensor in presence of 2.04 per cent of both the types of bacteria in water.

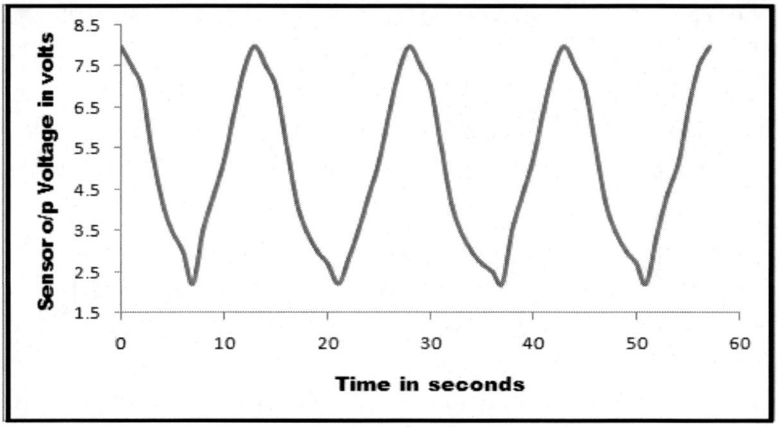

Figure 2.7. Repeatability test of the sensor for *E. coli* bacteria [37].

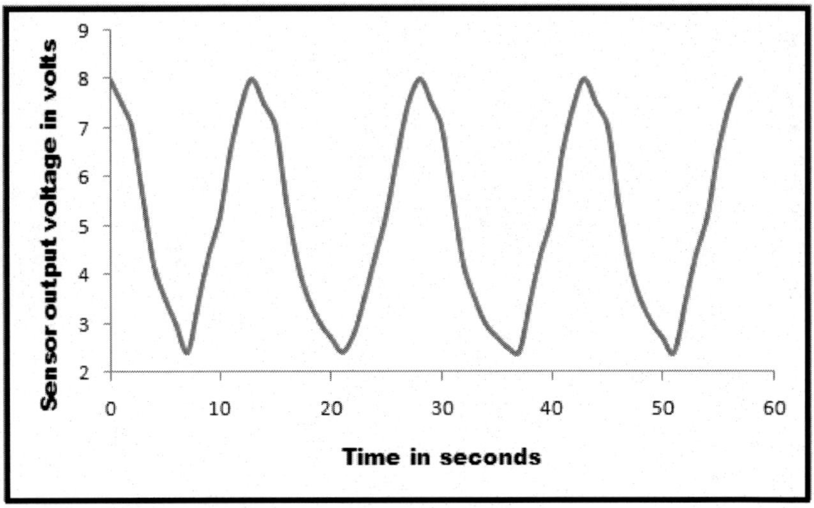

Figure 2.8. Repeatability test of the sensor for *S. pneumonia* bacteria [37].

2.3. Heavy Metal Detection

Metal ions are a major source of water contamination and this has encouraged researchers to develop novel low cost metal ion sensors to detect their presence. Although some of the heavy metals such as copper (Cu), iron (Fe) and zinc (Zn) are biologically essential for living organisms; they can lead to toxicity at higher concentrations and can cause serious concerns to human health [38].

Marina *et al.* [39] synthesized nanostructures of ZnO in four different morphologies: nanorods, nanoneedles, nanotubes and nanoplates. To determine the uniqueness of adsorption for each morphology, a series of electrochemical measurements were carried out using these nanostructured ZnO coatings on the working electrodes, using aqueous solutions of $Pb(NO_3)_2$ and $Cd(NO_3)_2$ as analytes with different concentrations. It was found that the sensitivity of the resulting electrochemical sensors depends on the morphology of the ZnO nanostructures: the best results were achieved in the case of porous nanostructures (nanotubes and nanoplates), whereas the lowest sensitivity corresponded to ZnO nanorods with a large diameter (i.e., low surface-to-volume ratio). It has been observed that ZnO morphologies exhibited significantly higher sensitivity in detecting lead ions compared to cadmium ions.

Borgohain *et al.*[40] fabricated UV detector by growing ZnO nanorods on Cu electrode. These UV detectors were used for detecting the concentration of heavy metals present in water samples. Initially, Chitosan capped ZnS quantum dots were added to the water samples mixed with heavy metals and as Chitosan is a very good metal chelating agent, ZnS quantum dots got aggregated. As a result of this aggregation, small clusters of ZnS quantum dots were formed. Depending on the concentration of metal ions in the water samples, the size of the clusters changes. When UV light was allowed to pass through the water samples and fall on the ZnO electrode, current in the order of μA was obtained flowing through the electrode. The absorption of UV light by the water samples varied with the variation in cluster sizes and therefore, current through the electrode also varied accordingly. The detector observations are shown in figure 2.9. (a) and figure 2.9. (b) with the variation in the observations was found to be within ± 5%.

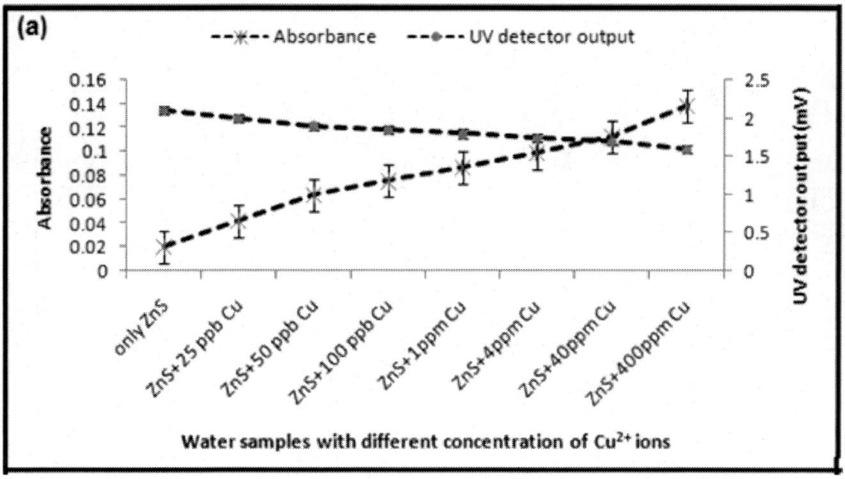

Figure 2.9. (a) Results of the UV detector in presence of Cu^{2+} ions [40].

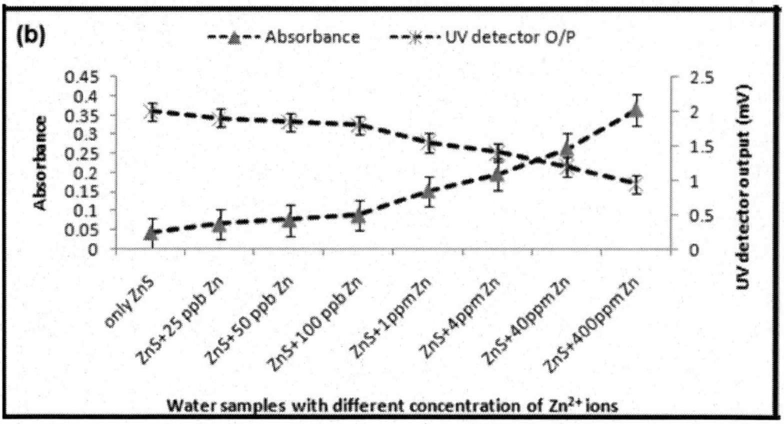

Figure 2.9. (b) Results of the UV detector in presence of Zn^{2+} ions [40].

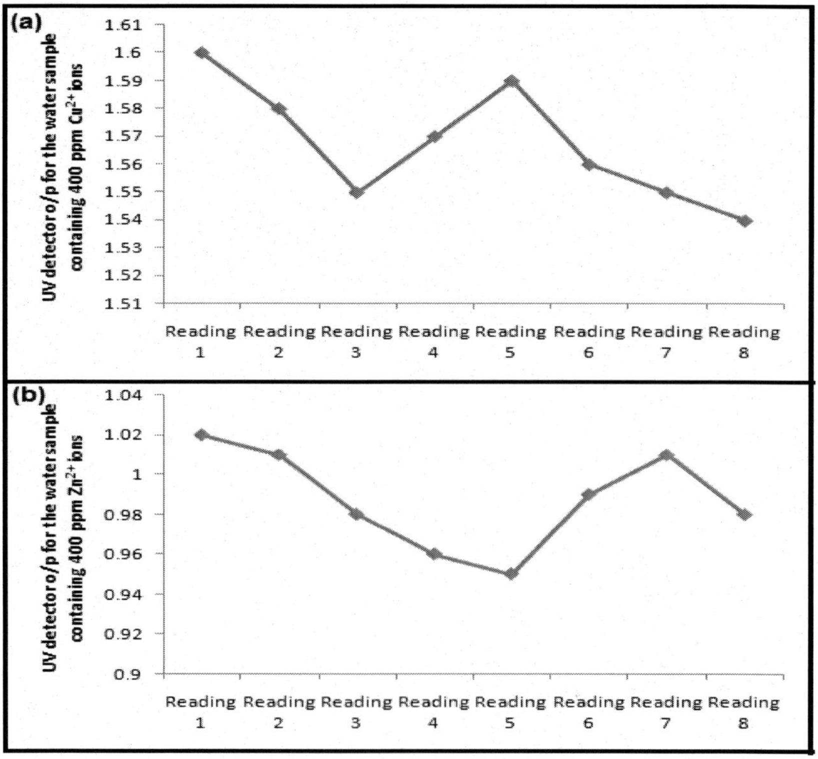

Figure 2.10. Repeatability of the sensor for (a) 400 ppm of Cu^{2+} ions (b) 400 ppm of Zn^{2+} ions [40].

The reproducibility and repeatability of the sensor was tested by taking several readings for the same water samples containing 400 ppm of Zn^{2+} ions and 400 ppm of Cu^{2+} ions. In figure 2.10. (a), it is seen that, the UV detector output for 400 ppm of Cu^{2+} ions has an average value of 1.57mV with a deviation within ±1.91 per cent and similarly, in figure 2.10. (b), it is seen that the UV detector output has an average value of 0.985mV in presence of 400 ppm of Zn^{2+} ions with a deviation within ±3.55 per cent.

3. Environmental Remediation

The past activities of human without consideration the impact on environment have harmed the environment in many ways. Various activities were carried out under numerous situations without any proper policy formulation or any act of law and regulations. Recently increased in new and effective nano scale materials have open door to many environmental applications. For example the use of nano scale materials to remediate contaminated surface and ground water at various sites is becoming one of the booming research areas. The nano scale materials are becoming choice of interest because of surface to volume ratio when compared with their volumes; therefore, their surface mediated reactions can be greatly enhanced in comparison to the same material at much larger sizes. The novel properties of these nanoscale materials can be used and manipulated for various specific applications which are not there within the micro- or macro scale counterpart. These novel properties allow better sensitivity and detection of contaminants and further can rapidly reduce the contaminants' concentrations. Even after all the ongoing research many applications of nanoscale materials for environmental remediation are in the developing phase and some are rapidly progressing from pilot-scale to full-scale implementation.

3.1. ZnO Nanostructures as Photo Catalytic Agent

Safe drinking water is the concern of the world since decades. To provide the same at minimum cost is the challenge. The requirement of new technologies for developing efficient point of use water treatment and other recycling systems for efficient disinfection and microbial control. In recent years, the use of photocatalytic process using semiconducting nanomaterials has shown substantial potential as a cost effective, eco-friendly and sustainable water treatment technology [41]. Several nanomaterials (both natural and engineered) have shown strong anti-microbial properties using different mechanism which also includes photocatalytic generation of reactive oxygen species damaging cell walls of bacteria and viruses. The ability of these nanomaterials including advanced oxidation possibilities has been widely utilized to remove harmful organic compounds and micro-organisms in water [42].

There are many conventional methods but heterogeneous photocatalysis techniques for water purification is gaining popularity as compared to other conventional methods because it does not generate harmful byproducts [43,44]. In this process the complex toxic long chain organic molecules are broken into benign fragments as well as the microbial cells are immobilized by rupturing the cell walls. In remote locations the problem of electricity is very common so the photocatalysis systems that utilizes sunlight is the interesting aspect which could allow energy efficient treatment. In semiconductors the light induced catalytic process of photocatalysis the photo-generated electron-hole pairs undergo redox reaction with molecules adsorbed onto the surface. This process further breaks the molecules into small fragments [45].

The mechanism which is followed in heterogeneous photocatalysis is diagrammatically shown in figure 3.1. The figure shows a wide bandgap semiconductor which acts as a photocatalytic material. This material when irradiated with light having energy higher than the band gap energy of the material, electron-hole pairs are generated. Material like TiO_2, ZnO, CdS, Fe_2O_3 and ZnS, which are semiconductor shows good sensitivity for redox processes which was induced by light due to the structure of atom, where there is an empty conduction band and a filled valence band [44]. In the process of generating electron hole pair, the electrons which are generated by the irradiation of photons moves up to the conduction band while the hole drifts to the bottom of the valence band. Most of the photo-generated electron-hole pairs undergo wasteful recombination, while the remaining initiate redox reactions in molecules which are adsorbed at the surface of the photocatalyst and thereby degrading them. The photo-generated electrons-holes pairs have been found to degrade almost all types of microbial and chemical contaminants [45].

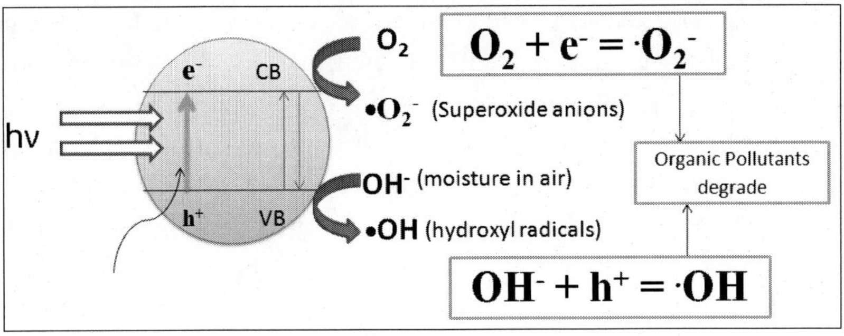

Figure 3.1. : Schematic diagram explaining photocatalysis [46].

The ZnO with a band gap of 3.37eV is one of the most popular and promising semiconducting material for photocatalysis [47]. Zinc oxide is listed as "generally recognized as safe" (GRAS) by the U.S. Food and Drug Administration. Nano-sized particles of ZnO have more effective anti-microbial activities than large particles, since the small sie leads to high surface to volume ratio of nano scale particles that allows better interaction with bacteria. Recent studies have shown that the toxicity of these nanoparticles on bacteria is very selective but exhibit minimal effects on human cells [48]. ZnO nanoparticles have been shown to exhibit antibacterial activity against a wide range of bacteria, both Gram negative bacteria and Gram positive bacteria, including major food borne pathogens like *Salmonella, Listeria monocytogenes*, *Staphylococcus aureus* and *Escherichiacoli*, [49]

The performance of a photocatalyst depends mainly on the ability to generate electron-hole pairs by absorbing electromagnetic waves, which contribute to photocatalysis through redox reactions [48]. Many research articles have shown that deionized water solvents Ethanol are used as solvent to synthesize zinc oxide nanoparticles using a simple chemical method. The optical absorption characteristics of the synthesized nano scale particles is determine to check how the nanoparticles can absorb maximum visible light in the optical band ranging from ultra violet to infrared bandwidth.

Figure 3.2. shows photocatalysis tests on using methylene blue (MB) as a test contaminant. The maximum (peak) absorption of methylene blueis at around 660 nm. The rapid degradation of methylene blue can be observed in the given graph visually. 10 μM MB in deionized water used along with the ZnO nanoparticles and exposed to visible light, which was not the case when it was exposed to sunlight without ZnO nanparticles. The degradation of methylene blue through redox process induced by photocatalysis using the synthesized nano scale zinc oxide particles is depicted by different graphs at different interval of time. It can be noted in the graph that the ZnO nanoparticles showed good photocatalytic activity by degrading the MB into benign form.

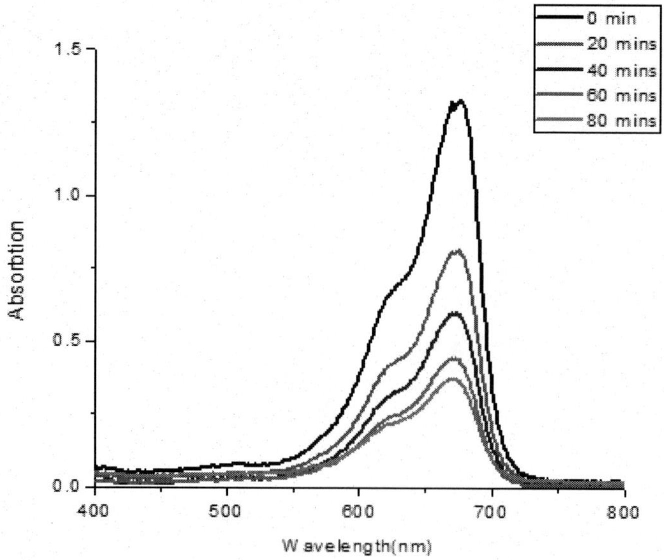

Figure 3.2. : Photocatalytic degradation of MB dye using ZnO nanoparticles [50].

There are several aspects, however, need to be taken into consideration for further development.

The use of ZnO nano particles as photocatalyst under suspended form gives very efficient result. But at the same time we need to be cautious about their applications. Depending on the application, the particles may have to retain separately. For example if the ZnO particles are used for destroying the microbial content in drinking water then the particles can't be left in suspended form in drinking water which no doubt will do the process of detoxifying the water. Instead we have to pass the contaminated drinking water through some media where the ZnO nanoparticles or other structures like rods are attached for photocatalysis and will not be suspended in drinking water.

Zinc oxide at nano scale is one of the most widely used standard photocatalysts in the field of environmental remediation. However, the large band gap and the massive recombination of photo catalytically generated electron-hole pairs in zinc oxide especially in its nanosize, limits the overall efficiency. This can be overcome by further modifying the electronic band structure of zinc oxide by hybridization or reengineering the same with a narrow band gap material, including metal, metal oxide, carbon based, and polymeric based. This process hybridization with the respective narrow band materials contribute to its sensitizer by shifting the absorption wavelength to the visible region of the spectrum [51].

3.2. ZnO Assisted CDI for Water Purification

Unavailability of drinking water is an alarming issue of global requirements and based on a recent report on water development from the United Nations, this crippling problem will only be increasing in the next 15 years or so. The current solution to these problems employing technologies to combat the water crisis includes desalination which has emerged as a main frame process to solve the drinking water issue worldwide. Available technology of desalination utilizes concepts like reverse osmosis (RO) and thermal processes. However a large amount of energy is required for these processes of desalination and also high maintenance costs. On the contrary, capacitive deionization (CDI) is a process which is membrane free and can be operated at low voltages which make it a god counterpart as energy efficient and low cost water desalination technique.

In 1960 Blair and Murphy reported the concept of electrochemical demineralization of water [52]. In 1968 the commercial relevance of CDI with it long term operation was demonstrated by Reid [53]. In 1970 theory was introduced by Johnson *et al.* which explains about the concept for the CDI process called 'potential modulated ion sorption', which is now well known as the Electric Double Layer (EDL) [54]. The research related to development of electrodes with more effective material took place from 1990 onward and then electrodes such as carbon aerogels or carbon nanotube electrodes attracted the CDI's attention [55]. The term "Capacitive Deionization" was introduced by Farmer et al. in 1996 which is now commonly abbreviation "CDI" [56]. The introduction of membrane in CDI was introduced in 2004 in a patent of Andelman [57].

By applying an electrical potential difference across two porous electrodes thereby creating an electric field in the process of Capacitive Deionization (CDI) technology is the key procedure to deionize water. Naturally anions which are negatively charged will be attracted towards the positive polarized electrode known as anodes and are removed from the water. Likewise, cations which are positively charged will be stored in the cathode, which is the negatively polarized electrode.

CDI is mostly used for brackish water treatment to desalinate. This water contains a low or moderate salt concentration (below 10 g/L). Distillation, reverse osmosis and electro dialysis are among the other technologies available for the deionization of water but CDI is a better option for brackish water treatment because of its energy efficient comapred to reverse osmosis and distillation. This is mainly because CDI removes the salt ions from the water, while the other technologies extract the water from the salt solution.

Figure 3.3. shows how CDI uses electrodes which are oppositely charged by applying a potential difference which in turn stores oppositely charged ions attracting them from the water. The arrangement of such electrodes can also be a group as assembled in stacks of multiple pairs. The contaminated water is allowed to pass through a ''spacer channel'' in between the two electrodes where the ions are harvested and are immobilized in the electrode porous surface. Formation of electrical double layers (EDLs) inside the intra particle pores is the basic mechanism of this process which salt ions are immobilized and selectively extracted from saline water. The process continues and a state comes when all of the accessible intra particle pore volume is saturated with adsorbed ions and the storage capacity of the device is reached. The elctrodes in this case have to be regenerated for further working. The adsorbed ions are released from the electrode by reversing the potential applied. In this a small stream of fresh water is allowed to pass through the space channel which collects the adsorbed ions on to the electrodes as the electrodes give away the ions because of opposite potential applied on them. So an enriched in ions water stream is produced and the electrodes regain their initial ion uptake capacity. This process of CDI is purely physical in nature and no chemical reaction happens in it. So it becomes a long term solution for desalination of water.

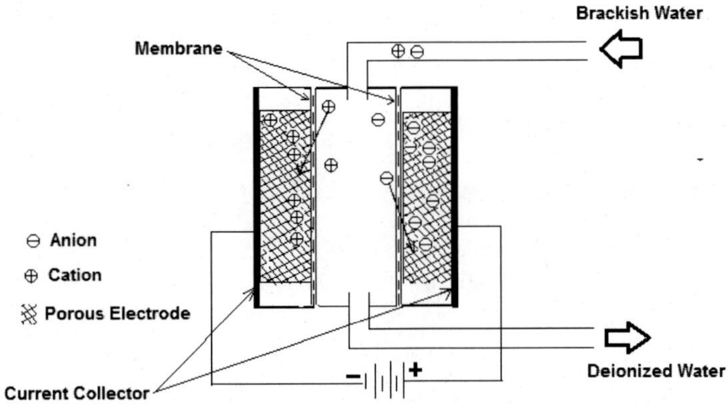

Figure 3.3. :-Working of the CDI cell.

Nanomaterials have shown research advancement for electrochemical desalination processes. The requirement of high surface area for the electrode used in CDI is becoming very essential for efficient collection of ions by warrant ion adsorption into the electrode [58].

This specific property of nanomaterial leads to large scale investment in nanotechnology for identifying effective electrodes enhanced with nano scale materials. So to enhance the surface area of the electrodes used in CDI,

ZnO at nano scale is considered to be very effective material for its large surface area availability for adsorption. Electrodes designed with activated carbon are most commonly used in CDI because of its low fabrication costs, inertness and large surface area. The process of ion adsorption is the function of the strength of electric field applied. On the other hand a considerable enhancement in the process of ion adsorption can be achieved by applying a coating of differently engineered ZnO nanostructure on the activated carbon electrode [59].

Zinc oxide (ZnO) has a wurtzitestructurein crystalline form, which leads to apreferential growth of nanorods in the vertical c-axis on awide array of surfaces [60]. Zinc oxide have shown a promising performance as amaterialin which the field emission is uniformly distributed along its surface[61].ZnO also because of its large surface to volume ratio can be used have effective space to absorb the ions. As the micopores and mesopores on to the surface of the activated carbon electrodes can be masked using ZnO nanorods but this in turn have so much of exposed nanorods surface that it increases the active surface area and the effective adsorption is better compared to the electrodes which are not coated with nanorods.

4. Clean Energy Generation

The world is at the moment largely dependent upon coal and fossil fuel. These forms of fuel are polluting and the reserves are also fast depleting. The need of the hour is smart energy solutions that deliver dependable and inexpensive electricity thereby contributing to a strong economy, without compromising on our health as well as that of the eco-system. No solitary energy technology can achieve all of these. The solution lies in a diverse energy approach that shows promise in moving the country towards a clean, sustainable energy future. Renewable energy is dependable, inexpensive, and advantageous for a healthy eco-system and a healthy economy. In this section, the use of nanostructured ZnO is discussed in two areas of clean energy generation: piezotronic energy generation and dye/quantum dot sensitized solar cells.

4.1. ZnO Based Piezo-Current Generator

Piezoelectricity is an inherent property of certain materials. These materials exhibit piezo-current instead of subjecting mechanical stress. Certain biological, crystalline, ceramic and polymeric materials exhibit this property [63]. Synthetic crystalline materials such as ZnO, CdS, GaN and ZnS of one-dimensional (1D) or two dimensional (2D) structures, are the prominent candidates for nanogenerator [64]. Among them 1D ZnO has been found to be the most sought after piezoelectric material as they manifest excellent elasticity, tensility and piezoelectricity [64, 65]. Wang *et al.* first introduce ZnO as a piezoelectric material [66]. The reported 1D ZnO structures as piezoelectric materials are nanorods (NRs) [67, 68], nanowires (NWs) [64] and nanobelts (NBs) [69]. The reported 2D ZnO structures are ZnO film [70] and ZnO-nanowalls [68]. 1D ZnO structures are generally more preferable for harvesting piezo-current in comparison to 2D ZnO. Fortunato *et al.* .reported that the piezoelectric response of 1D ZnO NRs is much better than 2D ZnO-nanowalls [68]. Normally vertical or lateral or tensile deforming force is applied on ZnO nanostructure to generate piezo-current. Vertical or lateral deforming force can be exerted on the ZnO nanowire (NW) by cantilever tip of atomic force microscopy (AFM) [64], scanning probe microscopy (SPM) [71] and peizoresponse force microscopy (PFM) [68]. Lateral deformation also can be possible through applying pressure or bending the ZnO NR grown on flexible substrate [72]. The tensile stress normally applied on ZnO thin film to get piezoresponse by stretching the ZnO coated flexible substrate.

4.1.1. Principle of Piezoelectric Behaviour in ZnO Nanostructures

Piezoelectric behaviour of ZnO occurs due to polarisation at atomic scale. The hexagonal structure of ZnO is basically a stack of alternating planes along c-axis. These planes are composed of one Zn^{2+} cation tetrahedrally surrounded by four O^{2-} anion. The centre of gravity (C.G) of the anion normally coincides with the cation which is at the centre of the tetrahedron. When a mechanical force is subjected to the ZnO crystal along the tetrahedron direction, the anions will refrain from coinciding with the same and thereby the tetrahedron experiences an electric dipole. If all such tetrahedron in the crystal does not take place in any dipole cancellation, the whole crystal would experience a macroscopic dipole. This gives opposite charges at two opposite faces [64, 66]. The whole phenomenon depict in the figure 4.1. (i).

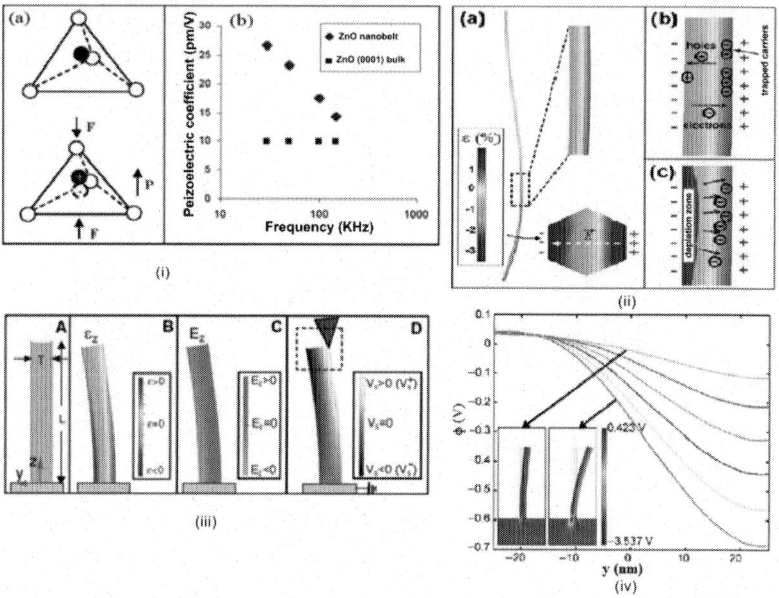

Figure 4.1. : (i) (a) Schematics depicts the dipole formation in tetrahedron in lieu of mechanical stress (b) The comparison of piezoelectric coefficient with varying frequency in case of ZnO bulk and ZnO nanobelt [66]. (ii) Schematics figure shows the responsible mechanism for conductance change (a) Strain distribution of the bent ZnO nanowire (using finite element simulation) (b) Trapping effect of the charge carriers (c) Formation of a charge depletion zone [64]. (iii) Piezo-potential developed in a single ZnO NR of length 1 μm and diameter 100 nm as considered by finite element simulation: (A) A vertically standing ZnO NR (B) ZnO NR deform vertically by the AFM tip, a strain εz is induced in the longitudinal direction (C) The developed electrical field Ez due to the mechanical deformation (D) Potential induced in the ZnO NR as a result of deformation [73]. (iv) Peizoelectric potential with varying applied force (F= 40 nN to F= 140 nN) [74].

A strain field is induced in the ZnO nanowire when it is vertically bent. The outer surface of the ZnO nanowire gets stretched and the inner surface compressed due to the bending of the ZnO NW. As a result of bending of ZnO NW a positive potential is generated at the stretched side and a negative potential at the compressed side (illustrated in figure 4.1. (ii)). The developed potential is because of relative displacement of the Zn^{2+} cation with respect to O^{2-} anion, which persists as long as strain remains. Similar phenomenon is observed in case of ZnO NR, when it deformed by a cantilever tip shown in figure 3 where the ZnO NR experiences positive (Vs+) potential at the stretched side and negative (Vs-) potential at the compressed side with respect to the base. This piezoelectric potential increases with increasing the applied force as shown in figure 4.1.(iv) [74].

4.1.2. Principle of Peizo-Current Harvesting Techniques

Wang *et al.* [73] used an AFM tip to deform and extract the induced charge from ZnO NR. However when this metal tip of AFM bend the n-type ZnO NR, the stretched side become positive and the compressed side become negative. The conduction band electron of n doped ZnO tends to accumulate at stretched (positive) side. These electrons thus create a screen at positive side while the potential at negative side preserves. As a result when this metal tip of AFM touches the stretched side of n-type ZnO NR, schottky barrier of reverse bias is formed which restrict the flow of current across the interface of metal-semiconductor and when this metal tip touches the compressed side, the schottky barrier shows forward bias as a result there is an electron flow from n-type ZnO NR to AFM tip which leads sudden increase in output current. In this process the extraction of charge was only possible from compressed side of ZnO NR. Wang *et al.* [75] introduced one another method to extract the charge from stretched side of ZnO NR, where they used p-type ZnO as a piezoelectric material. When this finite p-doped ZnO NR is laterally bent by AFM tip, the holes in p-type ZnO NR tend to accumulate at the compressed (negative) side. These holes partially create a screen at negative side while the positive potential at stretched side preserves. The stretched side of ZnO NR is initially slightly positive with respect to the ground of the NR. If the bending is increase further, the positive potential at stretched side become so strong which lead to drive the flow of charge carriers through the external circuit attached with the AFM tip. As a result a high positive voltage peak achieved from output. The minimum bending also can generate higher voltage peak in output if the p type doping is relatively very high. The comparative piezoelectric response of n-type and p-type ZnO NR is shown in figure 4.2.(i) and figure 4.2. (ii).

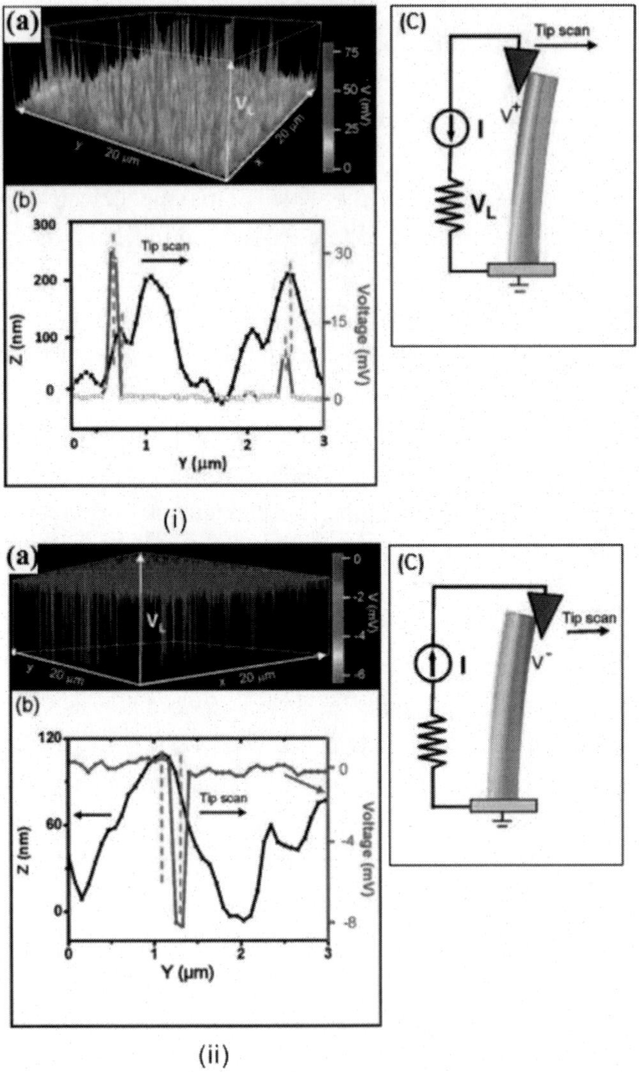

Figure 4.2. : (i) Piezoelectric response of n-type ZnO NR. (a) The recorded output voltage at the external load in three dimensional form when the AFM tip scanned over the NR arrays (b) A line scan profile from AFM topography (Black) and corresponding output voltage (Blue) images illustrated in (a). (c) Schematic of output voltage response for p-type ZnO NR [75]. (ii) Piezoelectric response of n-type ZnO NR. (a) The recorded output voltage at the external load in three dimensional form when the AFM tip scanned over NR arrays (b) A line scan profile from AFM topography (Black) and corresponding output voltage (Blue) images illustrated in (a). (c) Schematic of output response for n-type ZnO NR [75].

From the comparison of n-type and p-type ZnO NR, Wang *et al.* [75] reported that they find the average magnitude of positive voltage around 10-15 mV, and some pulses are as high as 50-90 mV for p-type ZnO NRs. The average magnitude of negative voltage around -5 to -10 mV found for n-type ZnO NRs. The output voltage peak position for n-type ZnO shows a delay in comparison to AFM topography profile while the peak position for p-type ZnO is synchronous with topography profile. This delay confirms that the voltage generate at output when the AFM tip touches the compressed side of n-type ZnO [75].

The current and power generated by a single nanowire or nanorod is not sufficient for a real device. A decent amount of piezoelectric voltage can be achieved by integrating a large no of nanowires. In this regards the main challenge is simultaneous extraction of piezoelectric response from the array of ZnO NWs. Researches modify the electrode to extract the induced charge simultaneously from each NW as well as modify the ZnO grown substrate to enhance the output voltage. Wang *et al.* [76] used a platinum coated silicon zig-zag electrode to extract the charge developed at ZnO NWs and to deform the corresponding ZnO NWs, ultrasonic vibration of 41 KH_Z were used. Qin *et al.* [77] use gold coated ZnO NW as an electrode. Xu *et al.* [78] reported that they have got excellent piezo voltage by using platinum coated flat surface as a electrode and gold coated substrate where ZnO NWs were grown. A poly (methyl methacrylate) or PMMA layer is used here to separate the electrode and the substrate. These array of ZnO NWs immersed in PMMA layer with exposing the tip. Choi *et al.* [72] design a mechanically powered Piezo-current generator using flexible material. ITO (indium tin oxide) coated flexible polyethersulfone or PES material used here as substrate and the palladium gold and ITO coated flexible material as a top electrode. ZnO NWs were used here as a piezoelectric material grown on the flexible substrate. There are various other possible methods to extract the induced charge viz gold deposited nanoporous AAO (Anodise aluminium oxide) template based electrode, self-assembledgold microwires based electrode using biological template etc [79]. Apart from ZnO nanostructure as a piezo-current generator, the piezoelectric property of this material have numerous applications viz field effect transistor (FET) [64], pressure sensor [72], self powered wearable device [80], amongst others.

4.2. Solar Energy Harvesting

The domain of solar cell research, presently, has been dominated by three major types viz (i) dye sensitized solar cell (DSSC) [81] (ii) bulk heterojunction (BHJ) photovoltaic cell or organic photovoltaic cell & (iii) Quantum Dot solar cell (QDSC) [82]. Several unique features of semiconductor nanostructures, viz simple fabrication process, tunability of band gap leading to efficient light absorption & possibility for designing flexible solar panel makes semiconductor nanostructures highly preferable for solar energy harvesting.

Dye-sensitized solar cells (DSSC) have received a lot of attention from researchers over the last decade. They are being considered as potential replacement to the existing p–n junction solar cells [81, 83-85]. A typical DSSC consists of a nanostructured metal oxide semiconductor film, preferably of titanium dioxide (TiO_2), zinc oxide (ZnO) and tin dioxide (SnO_2) [81,86-89]. The semiconductor part in a conventional photovoltaic system generates photoelectrons as well as transport the generated charge. In a DSSC, the sensitizer dye molecules adsorbed on to the semiconductor thin film are photo-excited. The injection of photoexcited electrons from the dye into the conduction band of the semiconductor creates charge separation at the interface. The carriers moving into the conduction band of the semiconductor finally moves to the external circuit. The schematic diagram shown in Figure 4.3 explains the principle of operation of a DSSC [84]. In a typical PN junction solar cell, the course of generation of electron-hole pair (charge formation) and diffusion of electrons and holes (charge transfer) takes place in the same material (normally silicon). However in a DSSC, the two major processes happen in two dissimilar materials. Figure 4.4 presents a comparison of the processes of charge generation and transfer in a PN junction solar cell and a DSSC [89].

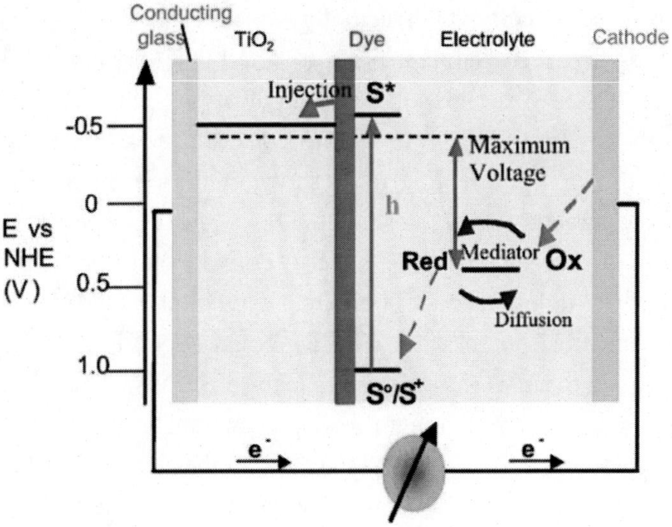

Figure 4.3.: Operating principle of the dye-sensitized solar cell. Photo-excitation of the sensitizer (S) leads to electron injection into the conduction band of a semiconducting metal oxide thin film. The dye molecule is replenished by the redox electrolyte. The redox system is regenerated at the counter-electrode by electrons passing through the external circuit. The difference between the redox potential of the mediator and the Fermi level of the semiconducting film gives a measure of the open circuit voltage of the DSSC.

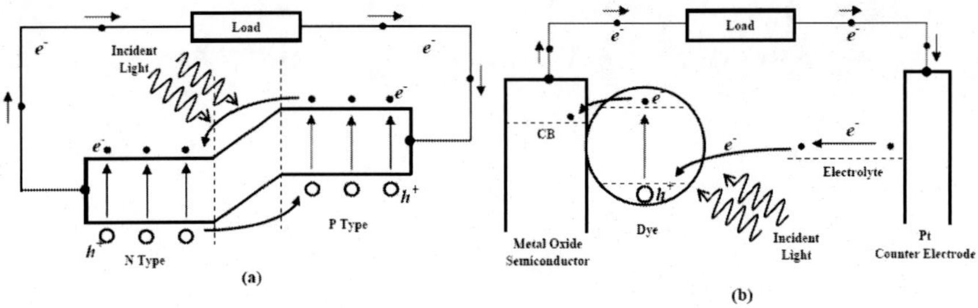

Figure 4.4.: Process of charge creation and charge transfer in (a) PN junction solar cell and (b) Dye sensitized solar cell (DSSC).

ZnO nanostructures are frequently used for solar cell application as it use for both viz as a transport layer for different sensitizers [82] and as a ballistic channel [82]. ZnO nanoparticles (NPs), ZnO microballs, ZnO nanowires, ZnO nanoflower etc. are different morphology of ZnO, used as a sensitizer in solar cell which actually generates electron-hole pair at incidence of sunlight. Baruah *et al.* [90] reported that they use N719 (dye)/ ZnO NP based sensitizer for solar cell application. They also use other nanostructures of ZnO with N719 dye to compare the energy conversion efficiency as shown in figure 4.5 (b). As shown in the schematic (Figure 4.5 (a)) the whole structure is immersed in I^-/I_3^- liquid redox electrolyte for transportation of charge carriers. The energy conversion is more in presence of UV light as UV light introduce an indirect excitation in N719 dye through the energy transfer process of FRET. When UV light incident on semiconductor, one part of the energy excite the electrons from valance band to conduction band and remaining part of energy transfer from excited state of semiconductor to the dye *via* FRET. Thereafter the electrons at the excited state of the dye move to the conduction band of the semiconductor and then easily move to collector. The dye N719 is attached with semiconductor due to surface absorption. The N719 (dye)/ ZnO NP based solar cell act as a regular DSSC in absence of UV light (illustrate in figure 4.5 (a)).

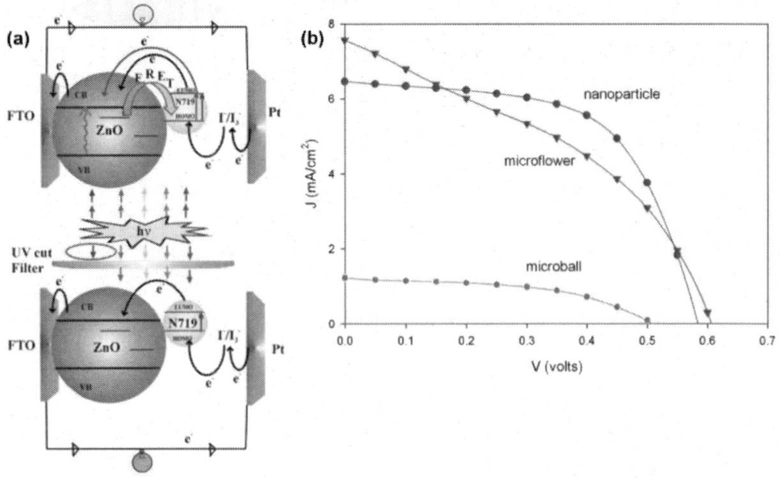

Figure 4.5.: (a) Schematic of sunlight conversion for ZnO NPs used DSSC. (b) I-V characteristics of DSSC using different ZnO nanostructures [90].

The Ballistic transport in ballistic channel attributes to transportation of charge carrier without any collision or resistance. One dimensional ZnO nanostructures i.e ZnO nanowire or nanorod are mostly used in any type of solar cell as a ballistic channel to avoid the electron hoping process in normal solar cell. These ZnO NWs also reduce the possibility of electron- hole recombination which actually enhance the conversion efficiency (illustrate in figure 4.6). Joel J. [82] reported that they found excellent improvement in short circuit current for ZnO NW as comparing to planar ZnO. They use ZnO/PbS base QD solar cell.

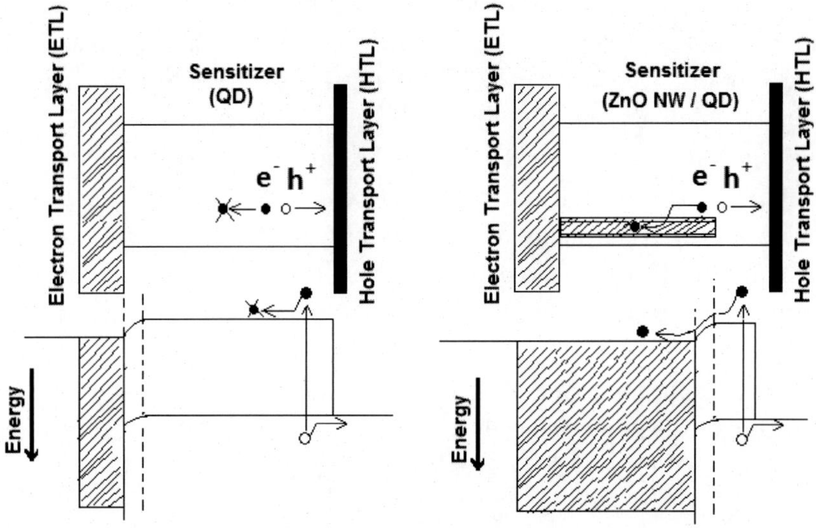

Figure 4.6.: (a) Schematic of ITO/ZnO seed layer /PbS QDs/MoO3/Au based solar cell. (b) Corresponding energy band diagram. (c) Schematic of ITO/ZnO seed layer/ ZnO NW /PbS QDs/MoO_3/Au based solar cell. (d) Corresponding energy band diagram. (e) Comparison of short circuit current for planar ZnO and ZnO NW [82].

5. Conclusion

ZnO is a technologically significant material and can be used for numerous applications like sensing, photocatalysis, energy harvesting, etc. In the nano regime, ZnO has the maximum reported morphologies like nanoparticles, nanowires, nanorods, nanoflowers, nanoprisms, nanohelixes, amongst others. The morphologies and the properties of nanostructured ZnO can be tailored through the control of growth parameters. Numerous reports are available in the literature on the use of nanostructured ZnO for environmentally friendly applications. Being a wide bandgap semiconductor (~3.36 eV) with a tunable bandgap, ZnO is ideal for sensing applications where electrons act as probes for different analytes. ZnO has been successfully used by researchers to detect trace level of both oxidizing and reducing gases as well as biological samples like gram positive and gram negative bacteria, molds, etc. Heterogenous photocatalysis has been proved to be an effective way of degrading harmful contaminants in water as well as immobilizing disease causing bacteria. ZnO is ideal for real field applications as defect sites can be incorporated within the bandgap, thereby allowing visible light absorption and this can tap the energy from a major part of the solar spectrum (almost 46%). The wurtzite structured ZnO, being piezotronic, can be used as a transducer to convert vibrational energy into electrical energy. With many researchers putting in their efforts into this domain, the new field of nano piezotronics evolved. ZnO thin films also act as very good transport layers for the new class of solar cells termed as dye/QD sensitized solar cells. Researchers have also shown that apart from acting as the transport layer, ZnO can also transfer a part of the photonic energy back to the dye molecules to generate extra electron hole pairs through a process called Forster Resonance Energy Transfer.

REFERENCES

[1] Baruah, S. and Dutta, J. (2009). Hydrothermal growth of ZnO nanostructures. *Sci. Technol. Adv. Mater.*, 10: 013001-013019.

[2] Zhong, L. W. (2004). Zinc oxide nanostructures: growth, properties and applications. *J. Phys.: Condens. Matter*, 16: 829–858.

[3] Baruah, S., Rafique, R. F. and Dutta, J. (2008). Visible light photocatalysis by tailoring crystal defects in ZnO nanostructures. *Nano*, 3 (V): 399–407.

[4] Guibal, E. (2004). Interactions of metal ions with chitosan-based sorbents: a Review *Sep. Purif. Technol.*, 38: 43–74.

[5] Baruah, S., Warad H.C., Chindaduang A., Tumcharern G. and Dutta J. (2008). Studies on Chitosan Stabilised ZnS: Mn2+ nanoparticles, *Journal of Bionanoscience*, 2: 1–7.

[6] Baruah, S., Ortinero, C. Shipin, O. V. and Dutta J., (2011).Manganese Doped Zinc Sulfide Quantum Dots for detection of Escherichia coli. *Journal of Fluorescence,* 22: 403–408.

[7] Baruah, S., Tumcharern G. and Dutta J. (2008). Chitosan Clad Manganese Doped Zing Sulphide Nanocrystallites for Biolabeling. *Advanced Materials Research,* 55-57: 589-592.

[8] Zhang, D., Liu, Z., Li, C., Tang, T., Liu, X., Han, S., Lei B. and Zhou, C. (2004). "Detection of NO2 down to ppb Levels Using Individual and Multiple In_2O_3 Nanowire Devices". *Nano Letters*, 4 (X): 1919-1924.

[9] Sivapunniyam, A., Wiromrat, N., Myint, M. T. Z. and Dutta, J. (2011). High-performance liquefied petroleum gas sensing based on nanostructures of zinc oxide and zinc stannate. *Sens. Actuators B: Chem.*, 157: 232-239.

[10] Hsueh, T. J., Hsu, C. L., Chang, S. J. and Chen, I. C. (2007). Laterally grown ZnO nanowire ethanol gas sensors. *Sensors and Actuators, B: Chemical,* 126 (II): 473-477.

[11] Sahay, P. P. (2005). Zinc oxide thin film gas sensor for detection of acetone, *Journal of Materials Science,* 40 (XVI): 4383-4385.

[12] Ippolito, S.J., Kandasamy, S., Kalantar-zadeh K. and Wlodarski W. (2005). Hydrogensensing characteristics of WO3 thin film conduct ometric sensorsactivated by Pt and Au catalysts. *Sens. Actuators B: Chem.*, 108: 154–158.

[13] Kolmakov, A., Klenov, D.O., Lilach, Y., Stemmer, S. and Moskovitst, M. (2005). Enhanced gas sensing by individual SnO2 nanowires and nanobelts functionalized with Pd catalyst particles, *Nano. Lett.,* 5: 667–673.

[14] Shen, Y., Yamazaki, T., Liu, Z., Meng, D. and Kikuta, T. (2009). Hydrogen sensing properties of Pd-doped SnO2 sputtered films with columnar nanostructures. *Thin Solid Films*, 517: 6119–6123.

[15] Lee, Y.I., Lee, K.J., Lee, D.H., Jeong, Y.K., Lee,H. S. and Choa, Y.H. (2009). Preparation and gas sensitivity of SnO2 nanopowder homogenously doped with Pt nanoparticles. *Curr. Appl. Phys.,* 9: 79–81.

[16] Downloaded from *http://www.uccee.org/Environmental_Pollution.html* on 28th March at 01:43 pm.

[17] Cabral, J. P. S. (2010). Water Microbiology. Bacterial Pathogens and Water. *International Journal of Environmental Research and Public Health, 7*: 3657-3703.

[18] Zhou, Q., Chen, W., Xu, L. and Peng, S. (2013). Hydrothermal synthesis of various hierarchical ZnO nanostructures and their Methane sensing properties. *Sensors*, 13: 6171-6182.

[19] Baruah, S. and Dutta, J. (2009). Effect of seeded substrates on hydrothermally grown ZnO nanorods. *J Sol-Gel Sci Technol*, 50: 456-464.

[20] Bai, S., Guo, T., Li, D., Luo, R., Chen, A. and Liu, C. C. (2013). Intrinsic sensing properties of the flower like nanostructures. *Sensors and Actuators B, 182*: 747-757.

[21] Yin, M., Liu, M. and Liu, S. (2013). Development of an alcohol sensor based on ZnO nanorods synthesized using a scalable solvothermal method. *Sensors and Actuators B, 185*: 735-742.

[22] Bai, S., Guo, T., Li, D., Luo, R., Chen, A. and Liu, C. C. (2013). Intrinsic sensing properties of the flower like nanostructures. *Sensors and Actuators B, 182*: 747-757.

[23] Pawar, R.C., Shaikh, J.S., Moholkar, A.V., Pawar, S.M., Kim, J.H., Patil, J.Y., Suryavanshi, S.S. and Patil, P.S. (2010). Surfactant assisted low temperature synthesis of nanocrystalline ZnO and its gas sensing properties. *Sensors and Actuators B,* 151: 212-218.

[24] Lu, G., Jing, X., Sun, J., Yu, Y., Zhang, Y. and Liu, F. (2012). UV enhanced room temperature NO_2 sensor using ZnO nanorods modified with SnO_2 nanoparticles. *Sensors and Actuators B, 162*: 82-88.

[25] Shouli, B., Liangyuan, C., Dianqing, L., Wensheng, Y., Pengcheng, Y., Zhiyong, L. Aifan, C. and Liu,C. C. (2010). Different

morphologies of ZnO nanorods and their sensing property. *Sensors and Actuators B, 146*: 129-137.

[26] Miller, D. R., Akbar, S. A. and Morris, P. A. (2014). Nanoscale metal oxide-based heterojunctions for gas sensing: A review. *Sensors and Actuators B*, 204: 250–272.

[27] Lu, G., Xu, J., Sun, J., Yu, Y., Zhang, Y. and Liu, F. (2012). UV enhanced room temperature NO_2 sensor using ZnO nanorods modified with SnO_2 nanoparticles. *Sensors and Actuators B,* 162: 82-88.

[28] Hwang, I. S., Kim, S. J., Choi, J. K., Choi, J., Ji, H., Kim, G. T., Cao, G. and Lee, J. H. (2010). Synthesis and gas sensing characteristics of highly crystalline ZnO–SnO_2 core–shell nanowires. *Sensors and Actuators B*, 148: 595–600.

[29] Park, S., An, S., Ko, H., Jin, C. and Lee, C. (2013). Enhanced NO2 sensing properties of Zn_2SnO_4-core/ZnO-shell nanorod sensors. *Ceramics International*, 39: 3539–3545.

[30] Borgohain, R., Das, R., Mondal, B., Yordsri, V., Thanachayanont, C. and Baruah, S. (2018). ZnO/ZnS Core-Shell Nanostructures for Low-Concentration NO2 Sensing at Room Temperature. *IEEE Sensors Journal*, 18 (XVII):7203-7208.

[31] Castillo, F. Y. R., Muro, A. L., Jacques, M., Garneau, P., González, F. J. A., Harel, J.and Barrera, A. L. G. (2015). Water borne Pathogens: Detection Methods and Challenges. *Pathogens,* 4: 307-334.

[32] Baron, E. J. (2011). Conventional versus Molecular Methods for Pathogen Detection and the Role of Clinical Microbiology in Infection Control. *Journal of Clinical Microbiology,* 49 (IX): 43.

[33] Roy, S. and Gao, Z. (2009). Nanostructure-based electrical biosensors. *Nanotoday,* 4 (IV): 318–334.

[34] Ivanova, I., Popova, R., Loukanov, A., Angelov, O., Krusteva, L., Papazova, K., Naumovska, E., Markoska, K. and Dushkin, C. (2010). Interaction between Bacillus Cereus and nanostructured thin films of Zinc oxide as a transducer element for Biosensing application. *Trakia Journal of Sciences*, 8 (II): 159-164.

[35] Tak, M., Gupta, V. and Tomar, M. (2014). Flower-like ZnO nanostructure based electrochemical DNA biosensor for bacterial meningitis detection. *Biosensors and Bioelectronics,* 59: 200–207.

[36] Tereshchenko, A., Fedorenko, V., Smyntyna, V., Konup, I., Konup, A., Eriksson, M., Yakimova, R., Ramanavicius, A., Balme, S. and Bechelany, M. (2016). ZnO Films Formed by Atomic Layer Deposition as an Optical Biosensor Platform for the Detection ofGrapevine Virus A-type Proteins. *Biosensors and Bioelectronics,* 92: 763-769.

[37] Borgohain, R. and Baruah, S. (2017). Development and Testing of ZnO Nanorods Based Biosensor on Model Gram-Positive and Gram-Negative Bacteria. *IEEE Sensors Journal*, 17 (IX): 2649-2653.

[38] Singh R., Gautam N., Mishra A. and Gupta R. (2011). Heavy metals and living systems: An overview. *Indian Journal of Pharmacology*, 43 (III): 246-253.

[39] Krasovska M., Gerbreders V. and Mihailova I. (2018). ZnO-nanostructure-based electrochemical sensor: Effect of nanostructure morphology on the sensing of heavy metal ions. *Beilstein Journal of Nanotechnology*, 9 (I): 2421-2431.

[40] Borgohain, R., Boruah, P. K. and Baruah, S. (2016). Heavy-metal ion sensor using chitosan capped ZnS quantum dots. *Sensors and Actuators B: Chemical*, 226: 534-539.

[41] Ong, C. B., Ng, L. Y. and Mohammad, A. W. (2018). A review of ZnO nanoparticles as solar photocatalysts: Synthesis, mechanisms and applications. *Renewable and Sustainable Energy Reviews*, 81:536-551.

[42] Lee,K. M., Lai,C. W., Ngai, K. S. and Juan,J. C. (2016). Recent developments of zinc oxide based photo catalyst in water treatment technology: A review. *Water Research*, 8: 428-448.

[43] Sugunan, A.and Dutta, J. (2008). Pollution treatment, remediation, and Sensing. Harald K (ed) *Nanotechnology*. Wiley-VCH, 9.

[44] Hoffmann, M. R. (1995). Environmental Applications of Semicondutor Photocatalysis. *Chemical Reviews,* 95: 69–96.

[45] Gaya, U.I. (2008). Heterogeneous photocatalytic degradation of organic contaminants over titanium dioxide: a review of fundamentals, progress and problems, *Journal of Photochemistry Photobiology C Photochemistry Reviews,* 9: 1–12.

[46] Baruah, S. and Dutta, J. (2009). Nanotechnology applications in pollution sensing and degradation in agriculture: a review. *Environ Chem Lett*, 7: 191–204.

[47] Hornyak, G.L. (2008). Introduction to nanoscience" *CRC Press, Boca Raton.*

[48] Sadollahkhani, A. (2014). Synthesis, structural characterization and photocatalytic application of ZnO@ZnS core–shell nanoparticles. *RSC Advances*, 4: 36940-36950.

[49] Kazeminezhad,I. (2014). Photocatalytic degradation of Eriochrome black-T dye using ZnO nanoparticles,. *Materials Letters*, 120: 267-270.

[50] Agarwal, B., Das, T. M. and Baruah, S. (2016). Improvement of photocatalytic activity of Zinc Oxide nanoparticles using Zinc Sulphide Shell. *ADBU-Journal of Engineering Technology*, 4: 137-141.

[51] Azam, M. A. M., Julkapli, N. M. and Hamid, S. B. A. (2016). Review on ZnO hybrid photocatalyst: impact on photocatalytic activities of water pollutant degradation. *Reviews in Inorganic Chemistry*, 36 (II): 1-28.

[52] Blair, J. W. and Murphy, G.W. (1960). Electrochemical demineralization of water with porous electrodes of large surface area. Saline watter conversion. *Advances in chemistry*, 27: 206–223.

[53] Reid, G.W. (1968).Field operation of 20 gallons per day pilot plant unit for electrochemical desalination of brackish water. *Washington D.C.: U.S. Dept. of the Interior*, 293.

[54] Johnson, A.M., Venolia, A.W., Wilbourne, R. G., Newman, J., Wong, C. M. and Gilliam, W.S. (1970). The electrosorb processes for desalting water. *Washington D.C.: U.S. Dept. of the Interior*, 516.

[55] Anderson, M.A., Cudero, A.L. and Palma, J. (2010)."Capacitive deionization as an electrochemical means of saving energy and

delivering clean water. Comparing to present desalination practices: Will it compete?. *Electrochimica Acta*, 55(XII): 3845-3856.

[56] Farmer, J.C., Fix,D.V., Mack, G.W., Pekala, R.W. and Poco, J.F. (1996).Capacitive deionization of NaCl and $NaNO_3$ solutions with carbon aerogel electrodes. *Journal of the Electrochemical Society*, 143: 159–169.

[57] Li, H., Gao, Y., Pan, L., Zhang, Y., Chen, Y. and Sun, Z. (2008).Electrosorptive desalination by carbon nanotubes and nanofibres electrodes and ion-exchange membranes. *Water Research*, 42: 4923-4928.

[58] Daer, S., Kharraz, J., Giwa, A. and Hasan, S. W. (2015). Recent applications of nanomaterials in water desalination: A critical review and future opportunities. *Desalination,* 367: 37-48.

[59] Laxman, K., Myint, M. T. Z., Khan, R., Pervez, T. and Dutta, J. (2015). Improved desalination by zinc oxide nanorod induced electric field enhancement in capacitive deionization of brackish water. *Desalination*, 359: 64-70.

[60] Baruah, S. and Dutta J. (2009). Hydrothermal growth of ZnO nanostructures. *Science and Technology of Advanced Materials*, 10: 1-18.

[61] Yu, K., Zhang, Y. S., Xu, F., Li, Q., Zhu, Z. Q. and Wan, Q. (2006). Significant Improvement of Field Emission by Depositing Zinc Oxide Nanostructures on Screen-Printed Carbon Nanotube Films. *Appl. Phys. Lett.*, 88: 153123-153126.

[62] Yan, C., Zou, L. and Short, R. (2014). Polyaniline-Modified Activated Carbon Electrodes for Capacitive Deionisation. *Desalination*, 333: 101−106.

[63] Skoog, D. A., Holler, F. J. and Crouch, S. R. (2007). *Principles of instrumental analysis*, 6th ed., Cengage learning.

[64] Wang, X.D., Zhou, J., Song, J.H., Liu, J., Xu, N.S. and Wang, Z.L. (2006). Piezoelectric field effect transistor and nanoforce sensor based on a single ZnO nanowire. *Nano Lett,*6 (XII):2768–2772.

[65] Xu F., Qin Q., Mishra A., Gu Y. and Zhu Y. (2010).Mechanical loading properties of ZnO nanowires under different loading modes. *Nano Res.*, 3 (IV): 271-280.

[66] Wang Z. L., Kong X. Y., Ding Y., Gao P., Hughes W. L., Yang R. and Zhang Y. (2004). Semiconducting and piezoelectric oxide nanostructures induced by polar surface. Adv. *Funct. Mater.,* 14 (X): 943-956.

[67] Hussain, M., Khan, A., Nur O., Willander, M. and Broitma E. (2014). The effect of oxygen-plasma treatment on the mechanical and piezoelectrical properties of zno nanorods. *Chemical Physics Letters*, 608: 235–238.

[68] Fortunato, M., Reddy, C. C., Bellis, G D., Ballirano, P., Soltani, P., Kaciulis, S., Caneve, L., Sarto, F. and Sarto, M. S.(2018). Piezoelectric Thin Films of ZnO-Nanorods/Nanowalls Grown by Chemical Bath Deposition. *IEEE Trans.,* 17 (II): 311-319.

[69] Zhao, M. H., Wang, Z. L. and Mao, S. X. (2004). Piezoelectric characterization of individual zinc oxide nanobelt probed by piezoresponse force microscope. *Nano Lett.,* 4 (IV): 587-590.

[70] Aleksandrova, M. Kolev G., Vucheva Y., Pathan H. and Denishev K.. (2017). Characterization of Piezoelectric Microgenerator with Nanobranched ZnO Grown on a Polymer Coated Flexible Substrate. *Appl. Sci., 7*: 890- 901.

[71] Agronin, A.G., Rosenwaks, Y. and Rosenman, G.I. (2003). Piezoelectric Coefficient Measurementsin Ferroelectric Single Crystals UsingHigh Voltage Atomic Force Microscopy. *Nano Lett.,* 3(II):169-171.

[72] Choi, M. Y., Choi, D., Jin, M. J., Kim, I., Kim, S. H., Choi, J. Y., Lee, Y. S., Kim, M. J. and Kim, S. W. (2009). Mechanically Powered Transparent flexible charge-generating nanodevices with piezoelectric ZnO nanorods. *Advanced Materials*, 21:2185-2189.

[73] Wang, Z. L., Song, J. H. Piezoelectric nanogenerators based on zinc oxide nanowire arrays. (2006). *Science,* 312:242-246.

[74] Wang, Z. L. (2008). Energy harvesting using piezoelectric nanowires - a correspondence on energy harvesting using nanowires. *Adv. Mater*, 20: 1-5.

[75] Lu, M.P., Song, J., Lu, M.Y., Chen, M.T., Gao, Y.,Chen, L. J. and Zhong, L. W. (2009). Piezoelectric Nanogenerator Using p-Type ZnO Nanowire Arrays. *Nano Letters*, 9 (III): 1223-1227.

[76] Wang, X., Song, J., Liu, J. and Wang, Z. L. (2007). Direct current nanogenerator driven by ultrasonic waves. *Science*, 316: 102-105.

[77] Qin, Y., Wang, X., and Wang, Z. L. (2008). Microfibre-nanowire hybrid structure for energy scavenging. *Nature,* 451 (IV): 809-813.

[78] Xu, S., Qin., Y., Xu, C., Wei, Y., Yang, R., and Wang, Z. L. (2010). Self-powered nanowire devices. *Nature Nanotechnology*, 5: 366-373.

[79] Sabah, A., Dauka, I., Kumar, P., Mohammed W. and Dutta, J. (2012). Growth of templated gold microwires by self organization of colloids on aspergillus niger. *Digest Journal of Nanomaterials and Biostructures*, 7 (II): 583-591.

[80] Park K., Jeong C. K., Kim N. K. and Lee K. J. (2016). Stretchable piezoelectric nanocomposite generator. *Nano Convergence*, 3 (XII): 1-12.

[81] Grätzel M.,(2009). Recent Advances in Sensitized Mesoscopic Solar Cells. *Acc. Chem. Res.,* 42(11):1788-1798.

[82] Jean, J., Chang, S., Brown, P. R., Cheng, J. J., Rekemeyer, P. H., Bawendi, M. G., et al. (2013). ZnO Nanowire Arrays for Enhanced Photocurrent in PbS Quantum Dot Solar Cells. *Advanced Materials,* 25 (XX): 2790–2796.

[83] Desjonqueres, M C and Spanjaard, D. (1993).Concepts in Surface Physics. *Springer Series in Surface Sciences.,* 30: (XVI) 607 p.

[84] Kim, Y. S., Ha, S. C., Kim, K., Yang, H., Choi, S. Y., Kim, Y. T., et al.(2005). Room-temperature semiconductor gas sensor based on nonstoichiometric tungsten oxide nanorod film. *Appl. Phys. Lett,* 86: 1-3.

[85] Yamazoe, N., Fuchigami, J., Kishikawa, M. and Seiyama, T. (1979). Interactions of tin oxide surface with O_2, H_2O and H_2. *Surf. Sci.*, 86: 335- 344.

[86] Wiromrat, N. (2008). *Master thesis AIT.*

[87] Gotz, A., Gracia, I., Plaza, J. A., Cane, C., Roetsch, P., Bottner, H. and Seibert, K. (2001). A novel methodology for the manufacturability of robust CMOSsemiconductor gas sensor arrays. *Sens. Actuators, B*, 77: 395-400.

[88] Chang, J. F., Kuo, H. H., Leu, I. C. and Hon, M. H. (2002). The Effect of Thickness and Operation Temperature of ZnO: Al Thin Film CO Gas Sensor. *Sens. Actuators, B*, 84: 258- 264.

[89] Xu, C., Tamaki, J., Miura, N. and Yamazoe, N. (1991). Grain size effects on gas sensitivity of porous SnO2-based elements. *Sens. Actuators B*, 3 (II): 147-155.

[90] Makhal, A., Sarkar, S., Bora, S., Baruah, S., Dutta, J., Raychaudhuri, A. K. and Pal, S. K. (2010). Role of Resonance Energy Transfer in Light Harvesting of Zinc Oxide-Based Dye-Sensitized Solar Cells. *J. Phys. Chem. C,* 114: 10390–10395.

BIOGRAPHICAL SKETCHES

Arnab Kumar Sarkar

Affiliation: Gauhati University

Education: B.Sc., M.Sc., PhD (pursuing).

Business Address: Department of Electronics and Telecommunication Engineering, School of Technology, Assam Don Bosco University, Airport road, Azara.

Research and Professional Experience:

Professional Experience:

A. Scientific Assistant

Dept of Electronics and Communication Engineering, Assam Don Bosco University, Guwahati.

March,2017 to till date.

B. Assistant professor (contractual)

Dept. of Physics, Abhayapuri college.

March, 2016 to May, 2016.

Research Experience:

A. Project Assistant

Dept of Electronics and Communication Technology, Gauhati University.

Funded by Assam Science Technology and Environment Council (ASTEC)

August 2016 to March 2017.

B. Working in the Nanotechnology Lab of Assam Don Bosco University in a collaborative project with Gauhati University on the development of low cost solar cell since September 2014 to January 2016.

Professional Appointments:

A. Scientific Assistant, Dept of Electronics and Communication Engineering, Assam Don Bosco University.

B. Project Assistant, Dept of Electronics and Communication Technology, Gauhati University.

C. Assistant Professor (contractual), Dept. of Physics, Abhayapuri College.

Honors: 1st Rank in M,Sc., Dept. of Electronics and Communication Technology, Gauhati University.

Publications from the Last 3 Years:

1. Rajbongshi H., Sarkar A.K., Phukan P, Bhattacharjee S. & Datta P., Ultrasensitive fluorescence detection of Fe^{3+} ions using fluorescein isothiocyanate functionalized $Ag/SiO_2/SiO_2$ core-shell nanocomposites, *Journal of Materials science: Materials in Electronics, Springer*; ISSN 0957-4522.
2. Sarkar A.K., Das H., Datta P., Absorption Enhancement of CdS nanoparticles using Poly Vinyl Alcohol (PVA), *Journal of Material Science and Mechanical Engineering (JMSME),* krishi Sanskriti Publications, Volume:3, Issue: 8, pp.526–528, p- ISSN: 2393-9095;e-ISSN: 2393-9109.

Bikash Agarwal

Affiliation: Assam Don Bosco University

Education: B.E., M.Tech., Ph.D (Pursuing)

Business Address: Department of Electronics and Telecommunication Engineering, School of Technology, Assam Don Bosco University, Azara, Guwahati 781017.

Research and Professional Experience: 10 Years

Professional Appointments:

- Senior Assistant Professor in the department of Electronics & Telecommunication Engineering, Assam Don Bosco University.
- Administrative Officer.

Honors: Awarded Gold Medal in M.Tech. degree.

Publications from the Last 3 Years:

1. S.K. Boruah, B. Agarwal, S. Baruah and P.K. Boruah "Nanostructured biosensor to estimate the freshness of fish" *ADBU Journal of Engineering Technology (AJET)*, Issue 1, Volume 7(2018), ISSN:2348-7305
2. B. Agarwal, P. Bhattacharyya, M. Goswami, D. Maiti, S. Baruah and P. Tribedi "Zinc oxide nanoparticle inhibits the biofilm formation of Streptococcus pneumoniae" *Antonie van Leeuwenhoek*, doi: 10.1007/s10482-017-0930-7 (2017), ISSN 0003-6072
3. B. Agarwal, T.M. Das and S. Baruah "Improvement of photocatalytic activity of Zinc Oxide nanoparticles using Zinc Sulphide shell" *ADBU Journal of Engineering Technology (AJET)*. Special Issue Volume 4(2016), ISSN:2348-7305

Sunandan Baruah

Affiliation: Assam Don Bosco University
Education: B.E., M.Tech., Ph.D
Business Address: Center of Excellence in Nanotechnology, School of Technology, Assam Don Bosco University, Azara, Guwahati 781017.

Research and Professional Experience: 20 Years

Professional Appointments:

- Director, Center of Excellence in Nanotechnology, Professor, Assam Don Bosco University.
- Professor and Head, Department of Electronics and Communication Engineering, Assam Don Bosco University.
- Visiting Faculty, Gauhati University, Assam
- Visiting Faculty, Asian Institute of Technology, Pathumthani, Thailand.

- Erasmus Mundus Post Doctoral Researcher, The ÅngströmLaboratoriet, Uppsala University, Sweden.
- Research Specialist**,** Centre of Excellence in Nanotechnology, Asian Institute of Technology, Thailand.
- Research Associate, Centre of Excellence in Nanotechnology, Asian Institute of Technology, Thailand.
- Lecturer, Electronics & Telecommunication, Residential Girls' Polytechnic, Golaghat, Assam, India.
- Lecturer, Assam Engineering College, Guwahati, Assam, India.
- Guest Lecturer, Department of Electronics, Gauhati University, Guwahati, Assam, India.

Honors: Awarded Gold Medal in M.Tech. degree.

Publications from the Last 3 Years:

1. K. Bhattacharyya, R. Deka, S. Baruah, "Automatic Target Recognition in RADAR at Terahertz Frequencies Using RNN and ANFIS: A Comparative Study", *International Journal of Modern Electronics and Communication Engineering* 7(2) (2019) 39-44.
2. K. Bhattacharyya, R. Thangjam, S. Goswami, K. Sarmah and S. Baruah, "Design and Analysis of Circular Slotted Microstrip Patch Antenna", *International Journal of Electronics and Telecommunications*, Polish Academy of Sciences, 65(3) (2019) 339–345.
3. K. Bhattacharyya, S. Goswami, K. Sarmah, S. Baruah, "A Linear-Scaling Technique for Designing a THz Antenna from a GHz Microstrip Antenna or Slot Antenna", *Optik-International Journal for Light and Electron Optics*, Elsevier, 199 (2019) 1-8. IF: 1.95.
4. K. Sarmah, A .Sarma, S. Goswami, S. Baruah, "Modified Groove Coupled Compact EBG Unit Cell as Notch Filter in a UWB Antenna", *International Journal of Electronics and Telecommunications*, Accepted for publication.

5. R. Borgohain, R. Das, B. Mondal, V. Yordsri, C. Thanachayanont and S. Baruah "ZnO/ZnS Core-Shell Nanostructures for Low-Concentration NO_2Sensing at Room Temperature" *IEEE Sensors Journal* 18(17) (2018) 7203 – 7208. *[Thompson Reuters Impact Factor: 2.512]*
6. S. K. Boruah, B. Agarwal, S. Baruah and P. K.Boruah "Nanostructured biosensor to estimate the freshness of fish" *ADBU Journal of Engineering Technology* 007010611 (5PP)
7. K. Das and S. Baruah "Quantum Dots for Solar Energy Harvesting: A Review" *Current Science* 115(4) (2018) 659-668. *[Thompson Reuters Impact Factor: 0.843]*
8. K. Bhattacharyya, R. Deka, S. Baruah "Automatic RADAR Target Recognition System at THz Frequency Band. A Review" *ADBU Journal of Engineering Technology* 6(3) (2017) 006031204 (15PP).
9. B. Sarkar, A. V. Daware, P. Gupta, K. K. Krishnani, S. Baruah, S. Bhattacharjee "Nanoscale wide-band semiconductors for photocatalytic remediation of aquatic pollution" *Environmental Science and Pollution Research* 24 (2017) 25775-25797. *[Thompson Reuters Impact Factor: 2.741]*
10. P. Bhattacharyya, B. Agarwal, M. Goswami, D. Maiti, S. Baruah and P. Tribedi "Zinc oxide nanoparticle inhibits the biofilm formation of *Streptococcus pneumonia" Antonie van Leeuwenhoek* 111 (1) (2018) 89-99. *[Thompson Reuters Impact Factor: 1.795]*
11. Imtinungla, A. Baruah, D. Lourembam and S. Baruah "Nanotechnology in Cancer Detection and Treatment" *ADBU Journal of Engineering Technology* 6(1) (2017) 00610607 (7pp)
12. C. S. Rahman, S. K. Boruah, P. K. Boruah, A. Borah, S. B. Dutta Borah and S. Baruah "Studies on the effect of temperature on Au Nanoparticles" *ADBU Journal of Engineering Technology* 6(1) (2017) 00610608(05pp).
13. R. Borgohain and S. Baruah "Development and Testing of ZnO Nanorods Based Biosensor on Model Gram-Positive and Gram-Negative Bacteria" *IEEE Sensors Journal* 17 (9) (2017), 2649-2653. *[Thompson Reuters Impact Factor: 2.512]*

14. A. Baruah, D. Lourembam, Imtinungla and S. Baruah"Nanotechnology for Water Purification" *ADBU Journal of Engineering Technology* 6(1) (2017) 00610603(06PP)
15. S. Goswami, K. Sarmah, A. Sarma, K. K. Sarma and S. Baruah "Design of a CSRR based Compact Microstrip Antenna for Image Rejection in RF-Downconverter based WLAN Receivers" AEU - *International Journal of Electronics & Communications* 74 (2017) 128-134. *[Thompson Reuters Impact Factor: 1.147]*
16. I. Dakua, J. K. Kasi, S. Baruah, N. Afzulpurkar "Novel electrode for tapping energy generated using piezotronic nanocrystals" *Journal of Nanoelectronics and Optoelectronics* 11(5)(2016) 631-637. *[Thompson Reuters Impact Factor: 0.497]*
17. R. Borgohain, S. Baruah "Design and analysis of UV detector using ZnO nanorods on interdigitated electrodes" *ADBU Journal of Engineering Technology* 4 (2016) 134-136.
18. B. Agarwal, T. M. Das, S. Baruah "Improvement of photocatalytic activity of ZnO nanoparticles using ZnS shell" *ADBU Journal of Engineering Technology* 4 (2016) 137-141.
19. D. Sarma, T. M. Das, S. Baruah "Bandgap Engineering of ZnO Nanostructures through hydrothermal growth" *ADBU Journal of Engineering Technology* 4 (2016) 216-218.

In: ZnO Nanostructures
Editor: Dana Crawford

ISBN: 978-1-53616-773-3

Chapter 2

ZnO NANOSTRUCTURES: AN ENVIRONMENTAL APPLICATION

***Mohammed Muzibur Rahman**, PhD**

Chemistry Department, Faculty of Science,
King Abdulaziz University, Jeddah, Saudi Arabia

ABSTRACT

In this approach, a large-scale synthesis of undoped low-dimensional semiconductor metal oxide nanostructures (ZnO nanoparticles, NPs) by simple wet-chemical method was performed using reducing agents at low temperature. The NPs were characterized in terms of their morphological, structural, and optical properties, and efficiently applied for the metal ions uptake. The detailed structural, compositional, and optical characterizations of the NPs were evaluated by powder X-ray diffraction pattern (XRD), Fourier-transform infra-red spectroscopy (FTIR), X-ray photoelectron spectroscopy (XPS), Electron dispersion spectroscopy (EDS), and UV-vis. spectroscopy, respectively which confirmed that the obtained NPs are well-crystalline undoped ZnO and possessed good optical properties. The ZnO NSs morphology was investigated by FESEM, which

* Corresponding Author's Email: mmrahman@kau.edu.sa.

confirmed that the calcined materials were spherical shape in nano-lavel and growth in huge-quantity. The analytical efficiency of newly synthesized ZnO NPs was also investigated for a selective separation of trivalent iron [Fe(III)] prior to its determination by inductively coupled plasma-optical emission spectrometry (ICP-OES). The selectivity of ZnO NPs towards different metal ions, including Cd(II), Co(II), Cr(III), Cu(II), Fe(III), Ni(II), Zn(II), and Zr(IV), was studied. Data obtained from the selectivity study suggested that that ZnO NPs phase was the most selective towards Fe(III). The static uptake capacity of Fe(III) was found to be ~79.80 mgg^{-1}. Moreover, adsorption isotherm data also provided that the adsorption process was mainly monolayer on a homogeneous adsorbent surface.

1. LITERATURE REVIEW

The development of simple, rapid and efficient methods has become of interest for monitoring metal ions in the environment. Several analytical methods have been applied to analyze metal ions in aqueous solutions, such as atomic absorption spectrometry [1], inductively coupled plasma-optical emission spectrometry (ICP-OES) [2], anodic stripping voltammetry [3], and ion chromatography [4]. However, analytical methods can not directly measure metal ions, in particular at ultra-trace concentration, in aqueous systems due to the lack of sensitivity and selectivity of these methods. Therefore, an efficient separation procedure is usually required prior to the determination of noble metals for sensitive, accurate and interference-free determination of noble metals [5]. Several analytical methods can be used for separation of analyte of interest, including liquid–liquid extraction [6], ion exchange [7], coprecipitation [8], cloud point extraction [9] and solid phase extraction (SPE) [10]. SPE is considered to be one of the most powerful techniques because it minimizes solvent usage and exposure, disposal costs, and extraction time for sample preparation. Several adsorbents have appeared because of the popularity of SPE for selective extraction of analytes, such as alumina [11], C18 [12], molecular imprinted polymers & cellulose [13], silica-gel [14, 15], activated carbon [16, 17] and carbon nanotubes [18, 19].

The semiconductor nanomaterials have attracted potential interest due to their unique properties and potential applications in all branches of science and technology. Semiconductor nanostructure materials have attracted an extensive attention owing to their unique properties and potential applications [21-22], which have been recognized as promising nanomaterials. It is exhibited a controlled morphology and composed of a number of irregular phases with geometrically-coordinated codoped metals and oxide atoms, which stacked alternately along the axes [23-25]. For characteristic physical and chemical properties of nanomaterials, ZnO semiconductor has been also explored the significant attention due to their large-surface area and controlled sizes [26-28]. However, ZnO nanomaterials have also drawn an enormous interest towards itself owing to their extrinsic, remarkable and wonderful features in electrical, optical, thermal, and mechanical properties as compared to their un-doping materials. It is also essential for the facile synthesis of ZnO nanoparticles in order to achieve the exceptional quality of semiconductor structures. Advances in nanotechnology with innovative solid crystalline phases, semiconductor nanomaterial have been regulating a key-task in the fabrication and improvement of very precise, perceptive, accurate, sensitive, and stable crystalline adsorbent. The exploration for even miniature nanomaterials accomplished of nano-level imaging and controlled morphology, the doping materials (even physical) have lately expanded the spot-light of awareness of the scientist, mainly for control determination, owing to the amplifying essential for environmental safety and health care fields [29, 30]. Semiconductor ZnO is the model materials for metal ions detection due to high crystalline surface areas and extensively employed as solid phase adsorbent for recognition, and quantification of various effective metal ions [31, 32].

Finally, the objective of this study was to investigate the analytical potential of ZnO NPs phase as an adsorbent on the selectivity and adsorption capacity of trivalent-iron [i.e., Fe(III)] prior to its determination by ICP-OES. The selectivity of ZnO NPs towards different metal ions, including Cd(II), Co(II), Cr(III), Cu(II), Fe(III), Ni(II), Zn(II) and Zr(IV), was evaluated in order to study the effectiveness of ZnO NPs on the adsorption of selected metal ions. Based on the selectivity study, it was concluded that the selectivity of ZnO NPs phase was the most towards Fe(III). The static uptake capacity for Fe(III) was determined to be ~79.80 mgg^{-1}. Results of adsorption isotherm also confirmed that the adsorption process was mainly monolayer on a homogeneous adsorbent surface. Adsorption data of Fe(III) were well fit with the Langmuir classical adsorption isotherm.

2. EXPERIMENTAL SECTIONS

2.1. Materials and Methods

Here, stock standard solutions of 1000.0 mgL^{-1} Cd(II), Co(II), Cr(III), Cu(II), Fe(III), Ni(II), Zn(II), and Zr(IV) were purchased from Sigma-Aldrich (Milwaukee, WI, USA; http://www.sigmaaldrich.com). Zinc chloride, sodium hydroxide, and other chemicals were used of analytical and spectral purity grade. Doubly distilled deionized water was also used throughout experimental studies. The powder X-ray diffraction (XRD) prototypes were assessed with X-ray diffractometer (Rigaku X-ray difractometer, Mini-Flex 2; http://www.rigaku.com) equipped with Cu-$K_{\alpha}1$ radiation (λ=1.5406 nm) using a generator voltage of 40.0 kV and a generator current of 35.0 mA applied for the purposed. The λ_{max} (377.0 nm) of calcined ZnO nanostructures was evaluated with UV/visible spectroscopy (UVO-2960, LABOMED Inc.; http://www.labomed.com). FT-IR spectra were performed with a spectrophotometer (Spectrum-100 FT-IR; http://www.perkinelmer.com) in the mid-IR range, which was purchased from Bruker, USA. The XPS measurements were executed for ZnO NPs by a Thermo Scientific K-Alpha (KA1066 spectrometer, Germany; http://www.thermoscientific.com). Monochromatic AlKα X-ray radiation sources were used as excitation sources, where beam-spot size was kept in 300.0 μm. The spectra was recorded in the fixed analyzer transmission mode, where pass energy was kept at 200.0 eV. The scanning of the spectra was performed at pressures less 10^{-8} Torr. Morphology of undoped ZnO

NPs was investigated on FESEM instrument (FESEM; JSM-7600F, Japan; http://www.jeol.co.jp). Elemental analysis of Zn NPs was investigated using EDS from JEOL, Japan. ICP-OES measurements were acquired by use of a Perkin Elmer ICP-OES (Model Optima 4100 DV, USA; http://www.perkinelmer.com). The ICP-OES instrument was optimized daily before measurement and operated as recommended by the manufacturers.

The ICP-OES spectrometer was used with following parameters: FR power, 1300 kW; frequency, 27.12 MHz; demountable quartz torch, Ar/Ar/Ar; plasma gas (Ar) flow, 15.0 $Lmin^{-1}$; auxiliary gas (Ar) flow, 0.2 $Lmin^{-1}$; nebulizer gas (Ar) flow, 0.8 $Lmin^{-1}$; nebulizer pressure, 2.4 bar; glass spray chamber according to Scott (Ryton), sample pump flow rate, 1.5 $mLmin^{-1}$; integration time, 3 s; replicates, 3; wavelength range of monochromator 165-460 nm. Selected metal ions were measured at wavelengths of 228.80 nm for Cd(II), 238.90 nm for Co(II), 267.72 nm for Cr(III), 327.39 nm for Cu(II), 259.94 nm for Fe(III), 221.65 nm for Ni(II), 206.20 nm for Zn(II) and 343.82 nm for Zr(IV).

2.2. Samples Preparation and Procedure for Metal Ions Uptake

Stock solutions of Cd(II), Co(II), Cr(III), Cu(II), Fe(III), Ni(II), Zn(II) and Zr(IV) were prepared in 18.2 MΩ·cm distilled deionized water and stored in the dark at 4.0 °C. For selectivity study, standard solutions of 5.0 mgL^{-1} of each metal ion were prepared and adjusted to pH value of 5.0 with acetate buffer. Then, each standard solution was individually mixed with 25.0 mg ZnO NPs. In this study, a fixed pH value of 5.0 was chosen for all metal ions in order to avoid any precipitation of other species, in particular for Fe(III). For example, Fe(III) usually forms a precipitation of $Fe(OH)_3$ with buffer solutions at pH value greater than 5.0. For the study of Fe(III) static adsorption capacity, standard solutions of 0, 5.0, 10.0, 15.0, 20.0, 25.0, 30.0, 50.0, 75.0, 125.0 and 150.0 mgL^{-1} were prepared as above, adjusted to the optimum pH value of 5.0 and individually mixed with 25.0 mg ZnO NPs. All mixtures were mechanically shaken for 1.0 hr at room temperature.

2.3. Preparation of ZnO NPs by Wet-Chemical Method

Initially zinc chloride were slowly dissolved into the de-ionized water to make 0.1 M concentration at room temperature ZnO nanoparticles have been prepared by adding zinc chloride precursors with active reducing agent into reactant beaker for 12.0 hours. Then these equi-mixtures were mixed gently and stirred until mix properly. The solution pH (>10) was slowly adjusted drop-wise by reducing agent (0.5M NaOH) solution. Then the mixture was placed into reactant cell to put on the hot-plate for longer time and maintained the temperature at 150.0°C. The starting materials of $ZnCl_2$, and NaOH were used without further purification for precipitation method to $Zn(OH)_2$ material. The excess NaOH was added dropwise into the vigorously stirred $ZnCl_2$ mixture to produce a white precipitate. The growth mechanism of the nanomaterials could be explained on the basis of chemical reactions and nucleation, as well as development of ZnO crystals. The probable reaction mechanisms are proposed for obtaining the undoped material oxides, which are presented in below.

$$NaOH_{(aq)} \rightarrow Na^+_{(aq)} + OH^-_{(aq)} \quad (i)$$

$$ZnCl_{2(s)} + 2OH^-_{(aq)} \rightarrow Zn(OH)_{2(aq)} + 2Cl^-_{(aq)} \quad (ii)$$

$$Zn(OH)_{2(aq)\ (heated)} \rightarrow ZnO_{(s)}\downarrow + H_2O \quad (iii)$$

The precursors of $ZnCl_2$ are soluble in DI and reacted in alkaline medium (0.5M NaOH reagent) according to the equation of (i) – (iii). After addition of NaOH into the mixture of metal oxides solution, it was stirred gently for few minutes (10.0 min) at room temperature. Then the solution pH is adjusted (pH>10) using NaOH and put into reactant cell to set onto hot-plate at 150.0°C for 12.0 hours, where the active temperature of the solution mixture is approximately in the range of 90.0 ~ 100.0°C. The reaction is progressed slowly according to the equation (iii). Then the solution was washed thoroughly with ethanol, acetone, and water and kept for drying at room temperature.

During the total synthesis process, NaOH acts a pH buffer to regulate the pH value of the solution and slow contribute of OH^- ions [33, 34]. When the concentration of the Zn^{2+} and OH^- ions is reached over the critical value, the precipitation of ZnO nuclei is initiated. As there is high concentration of Zn^{2+} ion in the solution, the nucleation of ZnO crystals become formed precipitation, due to the lower activation energy barrier of homogeneous nucleation. However, as the concentration of Zn^{2+} presences in the solution, some larger ZnO crystals with an aggregated morphology developed among the nanostructures, which composed of ZnO nanomaterials. Finally, the as-grown ZnO nanomaterials were calcined at 450.0°C for 5 hours in the furnace (Barnstead Thermolyne, 6000 Furnace, USA). In nanoparticles growth technique, initially ZnO nucleus growth takes place by self-aggregation, which then re-aggregates and formed ZnO nanocrystal according to the Ostwald ripening technique. Then nanostructure material crystallizes and re-aggregates with each other through Vander-Waals forces and produces ZnO spherical-shape NPs morphology, which is presented in Scheme 1. The calcined products were characterized in detail in terms of their morphological, structural, optical properties, and applied for metal ions detection.

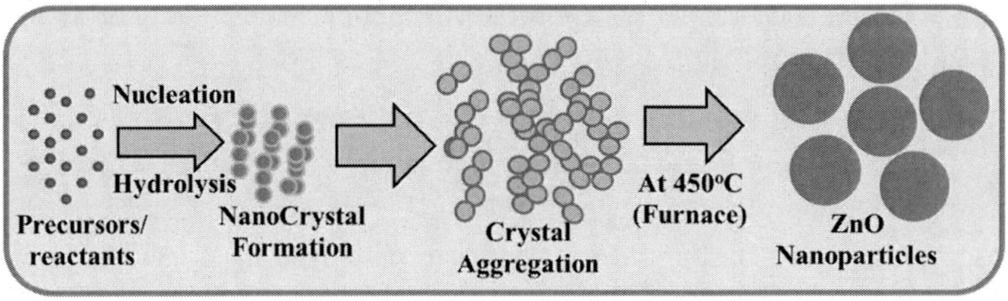

Scheme 1. Schematic representation of the growth mechanism of ZnO NPs by wet-chemical methods.

3. Results and Discussion

The undoped ZnO NP sample was analyzed and exhibited as wurtzite structure with hexagonal shapes. The sample was calcined at 450.0°C in furnace to ascertain the formation of nanocrystalline phases. Figure 1a shows typical crystallinity of the calcined ZnO nanoparticles and their aggregation. All the reflection peaks in this pattern were found to match with ZnO phase (zincite) having hexagonal geometry [Joint Committee on Powder Diffraction Standards, JCPDS # 071-6424]. The phases showed the major characteristic peaks with indices for calcined crystalline ZnO at 2θ values of 32.5(100), 34.9(002), 36.5(101), 47.4(102), 57.2(110), 63.6(103), and 68.7(200) degrees. The hexagonal (unit cell) lattice parameters are a = 3.2494, b = 5.2038, Point group: P63mc, and Radiation: CuKα1 (λ= 1.5406). These indicate that there is significant amount of crystalline ZnO present in nanoparticles.

The crystalline size was also calculated and confirmed using Scherrer formula (iv),

$$D = 0.9\lambda/(\beta\cos\theta) \quad \text{(iv)}$$

Where λ is the wavelength of x-ray radiation, β is the full-width at half maximum (FWHM) of the peaks at the diffracting angle θ. The average diameter of ZnO NPs is close to ~33.5 nm.

The optical absorption spectra of ZnO nanoparticle sample was investigated by using UV-vis. spectrophotometer in the visible range. The absorption spectrum of calcined undoped nanomaterials solution is presented in Figure 1b. It represents the absorption maxima at ~377.0 nm in visible range between 200.0 to 800.0 nm wavelengths, which indicated the formation of ZnO NPs formation by solution method route [35]. Band-gap energy is calculated on the basis of the maximum absorption band of ZnO NPs and obtained to be ~3.2891 eV, according to following equation (v).

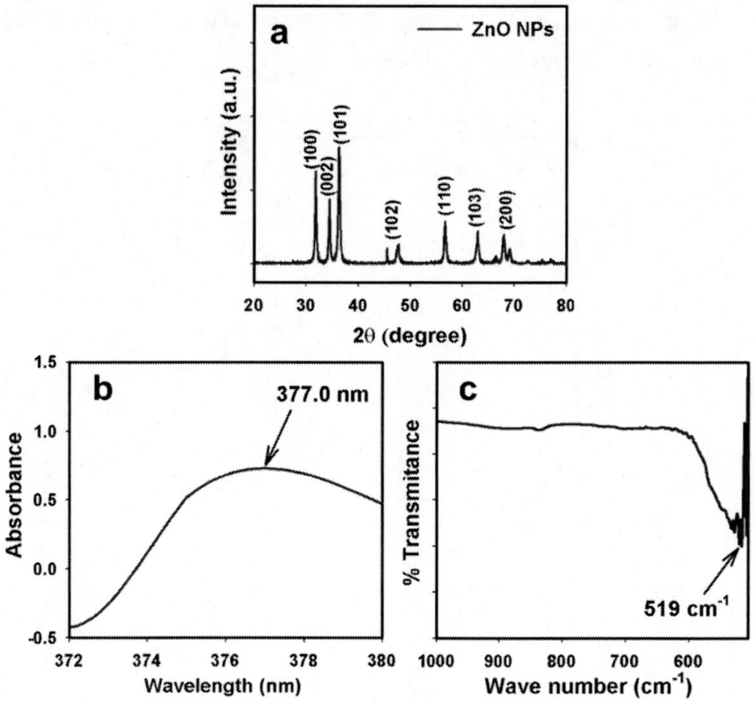

Figure 1. Evaluation of structural and optical properties. Typical (a) powder XRD, (b) UV/visible spectroscopy, and (c) FT-IR spectrums (% transmittance) of calcined ZnO NPs.

$$E_{bg} = \frac{1240}{\lambda} \text{ (eV)} \qquad \text{(v)}$$

Where E_{bg} is the band-gap energy and λ_{max} is the wavelength (377.0 nm) of the undoped ZnO NPs.

The calcined ZnO nanoparticle was also characterized from the atomic and molecular vibrations. To expect the activated recognition of nanomaterial clearly, FT-IR spectra only in the region of 500-1000 cm^{-1} were employed. Figure 1c represents the FT-IR spectrum of the calcined undoped ZnO nanoparticles. It exhibits a band at ~519.0 cm^{-1}. This observed vibration band may be assigned as metal-oxygen (Zn-O) stretching vibration. The observed vibration bands at low frequencies regions suggest the formation of ZnO nanomaterials.

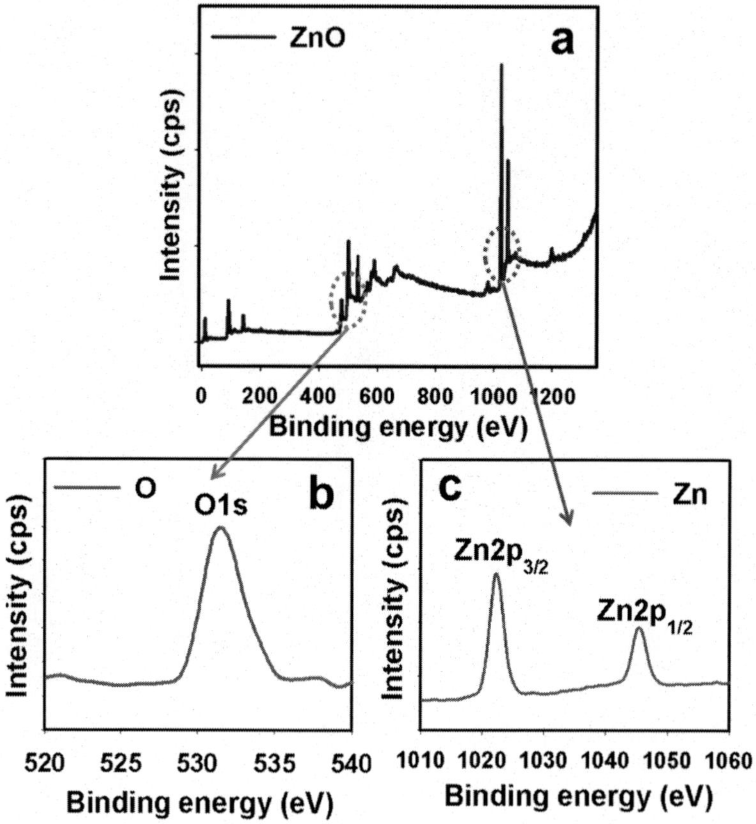

Figure 2. Evaluation of binding-energy of prepared nanomaterials. Typical XPS measurement of (a) ZnO NPs, (b) O1s level, and (c) Zn2p level acquired with MgKα1 radiations.

X-ray photoelectron spectroscopy (XPS) is a quantitative spectroscopic technique that investigated the elemental-composition, empirical-formula, chemical-state, and electronic-state of the elements that present within a nanomaterial. XPS spectra are acquired by irradiating a nanomaterial with a beam of X-rays, while simultaneously determining the kinetic energy and number of electrons that get-away from the top one to ten nm of the material being analyzed. Here, XPS measurements were investigated for undoped

ZnO nanoparticles to examine the chemical states of ZnO. The XPS spectra of Zn2p and O1s are presented in Figure 2a. The O1s spectrum shows a peak at 531.9 eV in Figure 2b. The peak at 531.9 eV is assigned to lattice oxygen, may be indicated to oxygen (ie, O_2^-) presence in the undoped ZnO nanoparticles [36]. In Figure 2c, the spin orbit peaks of the $Zn2p_{(3/2)}$ and $Zn2p_{(1/2)}$ binding energy for undoped nanoparticles appeared at around 1047.6 eV and 1024.8 eV respectively, which is in good agreement with the reference data for ZnO [37].

Figure 3. Evaluation of morphology of nanostructure materials. Typical (a-c) low to high-magnified FESEM images of ZnO NPs.

High resolution FESEM images of calcined ZnO NPs are presented in Figure 3(a-c). The images composed of nanostructure materials with particle shapes. In Figure 3(a-b), the diameter of ZnO NPs is calculated in the range of 25.0~55.0 nm, where the average value is close to ~34.0 ± 10.0 nm. It is clear from the FESEM images that the wet-chemically prepared products are nanostructure of ZnO, which propagated in a very high-density and possessed spherical nanoparticle shapes. Figure 2c represents the high-resolution (magnified) FESEM image of the calcined undoped NPs. It is reflected that the most of the nanostructure possessed in spherical-shapes of the aggregated undoped ZnO nanoparticles [38, 39].

Figure 4. Evaluation of elemental analysis of nanomaterials. Typical EDS measurement of undoped ZnO NPs including FESEM image, selected area of elemental analysis, and (inset) percentage of elements in terms of weight and atomic magnitudes.

The electron dispersive x-ray spectroscopy (EDS) analysis of these ZnO NPs indicates the presence of Zinc (Zn) and oxygen (O) composition in the pure calcined ZnO hetero-structures. It is clearly displayed that the calcined synthesize NPs contained only Zn and O elements, which is presented in Figure 4. No other peak related with any impurity has been detected in the FESEM coupled EDS (Inset of Figure 4), which confirms that the nanostructures are composed only with Zn (57.93%) and O (42.07%).

4. Application: Determination of Trivalent Iron Ions Uptake by Batch Method

4.1. Selectivity Study of ZnO NPs

Selectivity of the newly synthesized ZnO NPS phase towards different metal ions was examined based on determination of the distribution coefficient of ZnO NPs phase. The distribution coefficient (K_d) can be obtained from the following equation (vi) [40]:

$$K_d = (C_o - C_e / C_e) \times (V/m) \quad \text{(vi)}$$

where C_o and C_e refer to the initial and final concentrations before and after filtration with ZnO NPs, respectively, V is the volume (mL) and m is the weight of ZnO NPs phase (g). Distribution coefficient values of all metal ions investigated in this study are reported Table 1. It can be clearly noticed from Table 1 that the greatest distribution coefficient value was obtained for Fe(III) as compared to all metal ions. As shown in Table 1, the amount of Fe(III) was all extracted by ZnO NPs phase. Results of selectivity study provided that the newly synthesized ZnO NPs phase was the most selective towards Fe(III) among all metal ions included in this study. Thus, ZnO NPs phase is able to selectively bind with Fe(III), implying that the mechanism of adsorption may be electrostatic attraction or a chelating mechanism.

Table 1. Selectivity study of ZnO NPs phase adsorption towards different metal ions at pH 5.0 and 25.0°C (*N*= 5)

Metal Ion	q_e (mgg^{-1})	K_d (mLg^{-1})
Fe(III)	4.99	5.03032.26
Cr(III)	3.02	1523.98
Cu(II)	2.55	1038.32
Zn(II)	0.86	207.15
Ni(II)	0.73	170.69
Co(III)	0.62	141.29
Cd(II)	0.05	10.31
Zr(IV)	0.01	2.00

4.2. Static Adsorption Capacity

For determination of the static uptake capacity of Fe(III) on ZnO NPs phase, 25.0 mL Fe(III) sample solutions with different concentrations (0~150.0 mgL^{-1}) were adjusted to pH 5.0 and individually mixed with 25.0 mg ZnO NPs. These mixtures were mechanically shaken for 1.0 h at room temperature. Static adsorption capacity was obtained using equation (vii) as follows:

$$q_e = \frac{(C_o - C_e)V}{m} \quad \text{(vii)}$$

where q_e represents the adsorbed Fe(III) by the ZnO NPs phase (mgg^{-1}), C_o and C_e are the initial and equilibrium concentrations of Fe(III) ion in solution (mgL^{-1}), respectively, *V* is the volume (L) and *m* is the weight of ZnO NPs phase (g). Figure 5a and Figure 5b shows the static adsorption capacity and calibration curve (sensitivity study) of ZnO NPs for Fe(III) obtained from the experiment of adsorption isotherm respectively. In this study, the adsorption capacity of ZnO NPs for Fe(III) was determined to be 79.80 mgg^{-1}, which is comparable with those previously reported for Fe(III) in other studies (7.00 [41], 18.30 [42], 28.69 [43.], 28.90 [44], and 173.14 [45] mgg^{-1}).

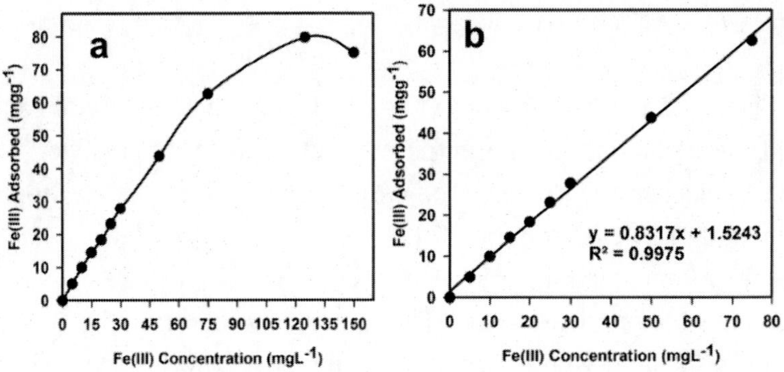

Figure 5. Study of metal ions uptake capacity. (a) Adsorption profile and (b) calibration curve (sensitivity calculation) of Fe(III) on 25.0 mg ZnO NPs phase in relation to the concentration at pH 5.0 and 25°C.

4.3. Adsorption Isotherm Models

Experimental equilibrium adsorption data were analyzed using different models in order to develop an equation that accurately represents the results. Langmuir equation is based on an assumption of a monolayer adsorption onto a completely homogeneous surface with a finite number of identical sites and a negligible interaction between the adsorbed molecules. The Langmuir adsorption isotherm model is governed by the following relation [46]:

$$C_e/q_e = (C_e/Q_o) + 1/Q_o b \quad \text{(viii)}$$

where C_e corresponds to the equilibrium concentrations of Fe(III) ion in solution ($mgmL^{-1}$) and q_e is the adsorbed metal ion by the adsorbate (mgg^{-1}). The symbols Q_o and b refer to Langmuir constants related to adsorption capacity (mgg^{-1}) and energy of adsorption (Lmg^{-1}), respectively. These constants can be determined from a linear plot of C_e/q_e against C_e with a slope and intercept equal to $1/Q_o$ and $1/Q_o b$, respectively. Moreover, the essential characteristics of Langmuir adsorption isotherm can be represented in terms of a dimensionless constant separation factor or equilibrium parameter, R_L, which is defined as [according to equation (ix)]

$$R_L = 1/(1 + bC_o) \quad \text{(ix)}$$

where b is the Langmuir constant (indicates the nature of adsorption and the shape of the isotherm); C_o the initial concentration of the analyte. The R_L value indicates the type of the isotherm, and R_L values between 0 and 1 represent a favorable adsorption [47].

The experimental isotherm data were fit well with the Langmuir equation based on the least square fit (Figure 6), supporting the validity of Langmuir adsorption isotherm model for the adsorption process similarly with the previous reports [48, 49]. Consequently, adsorption isotherm data indicated that the adsorption process was mainly monolayer on a homogeneous adsorbent surface. Langmuir constants Q_o and b are found to be ~79.63 mgg^{-1} and ~0.33 Lmg^{-1}, respectively. The correlation coefficient obtained from the Langmuir model is found to be R^2= 0.993 for adsorption of Fe(III) on ZnO NPs. Furthermore, the static adsorption capacity (~79.63 mgg^{-1}) calculated from Langmuir equation was in agreement with that (~79.80 mgg^{-1}) of the experimental isotherm study. The R_L value of Fe(III) adsorption on the ZnO NPs is 0.02, indicating a highly favorable adsorption process based on the Langmuir classical adsorption isotherm model.

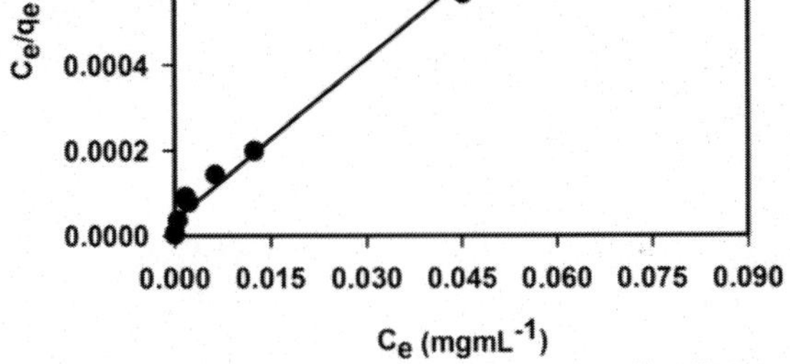

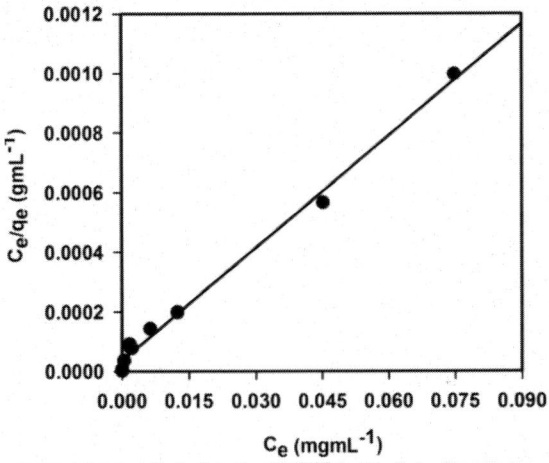

Figure 6. Study of Langmuir adsorption isotherm model. Study of trivalent iron, Fe(III) adsorption on 25.0 mg ZnO NPs phase at pH 5.0 and 25.0 °C. Adsorption experiments were obtained at different concentrations (0~150.0 mgL^{-1}) of Fe(III) under static conditions.

CONCLUSION

The present work offers a wet-chemically synthesize low-dimensional semiconductor metal oxide nanostructures (ZnO NPs), which was characterized using several conventional techniques like XRD, FE-SEM, XPS, EDS, UV/visible, and FT-IR spectroscopy etc. The detailed morphological characterizations by XRD and FESEM displayed that the synthesized NPs possess almost wurtzite structure with hexagonal spherical-shapes with typical diameters of ~33.5 nm. The optical properties of ZnO NPs were investigated by UV-visible absorption which shows the presence of characteristic band-gap energy of NPs peak ~3.2891 eV. The efficiency of the newly synthesized ZnO NPs phase for selective adsorption and determination of Fe(III) in aqueous solution was study by ICP-OES. Reasonable static adsorption capacity of ~79.80 mgg^{-1} with ZnO NPs adsorbent for Fe(III) in aqueous solution was obtained. Adsorption data of Fe(III) were well fit with the Langmuir adsorption isotherm model. Thus, the method may play an important role for using it as an effective approach for a selective adsorption and determination of tri-valent iron [Fe(III)] in complex matrices for healthcare and environmental fields.

REFERENCES

[1] S. Z. Mohammadi, D. Afzali, D. Pourtalebi, Flame atomic absorption spectrometric determination of trace amounts of lead, cadmium and nickel in different matrixes after solid phase extraction on modified multiwalled carbon nanotubes, *Cent. Eur. J. Chem.* 8 (2010) 662-668.

[2] H. J. Cho, S. W. Myung, Determination of cadmium, chromium and lead in polymers by icp-oes using a high pressure asher (hpa), Bull. *Korean Chem. Soc.* 32 (2011) 489-497.

[3] H. Xu, Q. Zheng, P. Yang, J. Liu, L. Jin, Sensitive voltammetric detection of trace heavy metals in real water using multi-wall carbon

nanotubes/nafion composite film electrode, *Chin. J. Chem.* 29 (2011) 805-812.

[4] S. Tanikkul, J. Jakmunee, S. Lapanantnoppakhun, M. Rayanakorn, P. Sooksamiti, R. E. Synovec, G. D. Christian, K. Grudpan, Flow injection invalve mini-column pretreatment combined with ion chromatography for cadmium, lead and zinc determination, *Talanta* 64 (2004) 1241-1246.

[5] K. Pyrzynska, Recent developments in the determination of gold by atomic spectrometry techniques, Spectrochim. *Acta Part B* 60 (2005) 1316-1322.

[6] A. Nasu, S. Yamaguchi, T. Sekine, Solvent extraction of copper(i) and (ii) as thiocyanate complexes with tetrabutylammonium ions into chloroform and with trioctylphosphine oxide into hexane, *Anal. Sci.* 13 (1997) 903-911.

[7] G. H. Tao, Z. Fang, Dual stage preconcentration system for flame atomic absorption spectrometry using flow injection on-line ion-exchange followed by solvent extraction, *J. Anal. Chem.* 360 (1998) 156-160.

[8] M. Soylak, N. D. Erdogan, Copper (ii)-rubeanic acid coprecipitation system for separation-preconcentration of trace metal ions in environmental samples for their flame atomic absorption spectrometric determinations, *J. Hazard. Mater.* 137 (2006) 1035-1041.

[9] J. L. Manzoori, H. Abdolmohammad-Zadeh, M. Amjadi, Simplified cloud point extraction for the preconcentration of ultra-trace amounts of gold prior to determination by electrothermal atomic absorption spectrometry, *Microchim. Acta* 159 (2007) 71-78.

[10] G. Fang, J. Chen, J. Wang, J. He, S. Wang, N-methylimidazolium ionic liquid-functionalized silica as a sorbent for selective solid-phase extraction of 12 sulfonylurea herbicides in environmental water and soil samples, *J. Chromatogr. A* 1217 (2010) 1567-1574.

[11] R. Ahmad, R. Kumar, Adsorption of amaranth dye onto alumina reinforced polystyrene, *Clean: Soil, Air, Water* 39 (2011) 74-82.

[12] S. Pei, Z. Fang, Flame atomic absorption spectrometric determination of silver in geological materials using a flow-injection system with on-line preconcentration by coprecipitation with diethyldithio-carbamate, *Anal. Chim. Acta* 294 (1994) 185-193.

[13] C. Gustavo Rocha de, A. Ilton Luiz de, R. Paulo dos Santos, Synthesis, characterization and determination of the metal ions adsorption capacity of cellulose modified with p-aminobenzoic groups, *J. Mater. Res.* 7 (2004) 329-334.

[14] K. Wei, L. Shu, W. Guo, Y. Wu, X. Zeng, Synthesis of amino-functionalized hexagonal mesoporous silica for adsorption of Pb^{2+}, *Chin. J. Chem.* 29 (2011) 143-146.,

[15] Y. Liu, L. Guo, L. Zhu, X. Sun, J. Chen, Removal of cr(iii, vi) by quaternary ammonium and quaternary phosphonium ionic liquids functionalized silica materials, *Chem. Eng. J.* 158 (2010) 108-114.

[16] H. M. Marwani, H. M. Albishri, T. A. Jalal, E. M. Soliman, Activated carbon immobilized dithizone phase for selective adsorption and determination of gold(iii), *Desalin. Water Treat.* 45 (2012) 128-135.

[17] H. M. Marwani, H. M. Albishri, E. M. Soliman, T. A. Jalal, Selective adsorption and determination of hexavalent chromium in water samples by chemically modified activated carbon with tris(hydroxymethyl)aminomethane, *J. Disp. Sci. Technol.* 33 (2012) 549-555.

[18] S. Tong, S. Zhao, W. Zhou, R. Li, Q. Jia, Modification of multi-walled carbon nanotubes with tannic acid for the adsorption of La, Tb and Lu ions, *Microchim. Acta* 174 (2011) 257-264.

[19] P. Biparva, M. R. Hadjmohammadi, Selective separation/ preconcentration of silver ion in water by multiwalled carbon nanotubes microcolumn as a sorbent, *Clean: Soil, Air, Water* 39 (2011) 1081-1086.

[20] M. Yoshida, J. Lahann. Smart nanomaterials, *ACS Nano.* 2 (2008) 1101.

[21] A. Umar, M. M. Rahman, S. H. Kim, Y. B. Hahn. Zinc oxide nanonail based chemical sensor for hydrazine detection, *Chem. Commun.* (2008) 166.

[22] D. Feng, W. Luo, J. Zhang, M. Xu, R. Zhang, H. Wu, Y. Lv, A. M. Asiri, S. B. Khan, M. M. Rahman, G. Zheng, D. Zhao. Multi-layered Mesoporous TiO_2 Thin films with large pore and highly crystalline frameworks for efficient photoelectrochemical conversion, *J. Mater. Chem.* A 1 (2013) 1591.

[23] Willner, B. Willner. Biomolecule-based nanomaterials and Nanostructures, *Nano let.* 10 (2010) 3805.

[24] M. M. Rahman, S. B. Khan, H. M. Marwani, A. M. Asiri. K. A. Alamry. Selective Iron(III) ion uptake using CuO-TiO_2 nanostructure by inductively coupled plasma-optical emission spectrometry, *Chem. Cent.* J. 6 (2012) 158.

[25] M. Tan, P. Munusamy, V. Mahalingam, F. C. J. M. Van-Veggel. Blue electroluminescence from InN@SiO_2 Nanomaterials, *J. Am. Chem. Soc.* 129 (2007) 14122.

[26] K. Ghosh, M. Kanapathipillai, N. Korin, J. R. McCarthy, D. E. Ingber. Polymeric Nanomaterials for Islet targeting and Immunotherapeutic deliver, *Nano Lett.* 12 (2012) 203.

[27] M. M. Rahman, A. Jamal, S. B. Khan, M. Faisal. Fabrication of Highly Sensitive Ethanol Chemical Sensor Based on Sm-Doped Co_3O_4 Nano-Kernel by Solution Method, *J. Phys. Chem. C* 115 (2011) 9503.

[28] M. M. Rahman, G. Gruner, M. S. Al-Ghamdi, M. A. Daous, S. B. Khan, A. M. Asiri. Chemo-sensors development based on low-dimensional codoped α-Mn_2O_3-ZnO Nanoparticles using flat-silver electrodes, *Chem. Cent. J.* 7 (2013) 60.

[29] H. Liu, Y. Li, S. Xiao, H. Gan, T. Jiu, H. Li, L. Jiang, D. Zhu, D. Yu, B. Xiang, Y. Chen. Synthesis of Organic One-Dimensional Nanomaterials by Solid-Phase Reaction, *J. Am. Chem. Soc.*, 125 (2003) 10794.

[30] M. M. Rahman, S. B. Khan, H. M. Marwani, A. M. Asiri, K. A. Alamry, A. O. Al-Youbi. Selective determination of gold(III) ion using CuO microsheets as a solid phase adsorbent prior to ICP-OES measurements, *Talanta* 104 (2013) 75.

[31] B. Rybtchinski. Adaptive Supermolecular nanomaterials based on strong noncovalent interactions, *ACS Nano* 5 (2011) 6791.

[32] M. M. Rahman, A. Umar, K. Sawada. Development of Amperometric Glucose Biosensor Based on Glucose Oxidase Enzyme Immobilized with Multi-Walled Carbon Nanotubes at Low Potential, *Sens. Actuator: B Chem.* 137 (2009) 327.

[33] M. Darroudi, Z. Sabouri, R. K. Oskuee, A. K. Zak, H. Kargar, M. H. N. A. Hamid. Green chemistry approach for the synthesis of ZnO nanopowders and their cytotoxic effects, *Ceramic Internation* 40(2014)4827-4831.

[34] M. M. Rahman, S. B. Khan, H. M. Marwani, A. M. Asiri, K. A. Alamry, A. O. Al-Youbi. Selective determination of gold(III) ion using CuO microsheets as a solid phase adsorbent prior by ICP-OES measurement, *Talanta* 104(2013)75-82.

[35] Y. Liu, Y. Deng, Z. Sun, J. Wei, G. Zheng, A. M. Asiri, S. B. Khan, M. M. Rahman, D. Zhao. Hierarchical Cu_2S Microsponges Constructed from Nanosheets for Efficient Photocatalysis, *Small* 9(2013)2702-2708.

[36] M. M. Rahman, S. B. Khan, M. Faisal. M. A. Rub, A. O. Al-Youbi, A. M. Asiri. Determination of Olmisartan medoxomil using hydrothermally prepared nanoparticles composed SnO_2-Co_3O_4 nanocubes in tablet dosage forms, *Talanta* 99 (2012) 924-931.

[37] AEnnaoui, J. Klaer, T. Kropp, R. Saez-Araoz, N. Allsop, I. Lauermann, H. W. Schock, M. C. Lux-Steiner. Formation of a ZnS/Zn(S,O) bilayer buffer on CuInS2 thin film solar cell absorbers by chemical bath deposition, *J. App. Phys* 99 (2006)1235803.

[38] I. T. Liu, M. H. Hon, L. G. Teoh. The preparation, characterization and photocatalytic activity of radical-shaped CeO_2/ZnO microstructures, *Ceramics International* 40(2014)4019-4024.

[39] G. Srinet, R. Kumar, V. Sajal. Effect of aluminium doping on structural and photoluminescence properties of ZnO nanoparticles, *Ceramics International* 40(2014)4025-4031.

[40] D. M. Han, G. Z. Fang, X. P. Yan, Preparation and evaluation of a molecularly imprinted sol-gel material for on-line solid-phase extraction coupled with high performance liquid chromatography for

the determination of trace pentachlorophenol in water samples, *J. Chromatogr. A* 1100 (2005) 131-136.

[41] D. W. O'Connell, C. Birkinshaw T. F. O'Dwyer, Heavy metal adsorbents prepared from the modification of cellulose: A review, *Bioresour. Technol.* 99 (2008) 6709-6724.

[42] S. Tokalioglu, V. Yilmaz, S. Kartal, A. Delibas, C. Soykan, Synthesis of a novel chelating resin and its use for selective separation and preconcentration of some trace metals in water samples, *J. Hazard. Mater.* 169 (2009) 593–598.

[43] Z. Zang, Z. Hu, Z. Li, Q. He, X. Chang, Synthesis, characterization and application of ethylenediamine-modified multiwalled carbon nanotubes for selective solid-phase extraction and preconcentration of metal ions, *J. Hazard. Mater.* 172 (2009) 958-963.

[44] Z. Li, X. Chang, X. Zou, X. Zhu, R. Nie, Z. Hu, R. Li, Chemically-modified activated carbon with ethylenediamine for selective solid-phase extraction and preconcentration of metal ions, *Anal. Chim. Acta* 632 (2009) 272-277.

[45] K. N. Ghimire, J. Inoue, K. Inoue, H. Kawakita, K. Ohto, Adsorptive separation of metal ions onto phosphorylated orange waste, *Sep. Sci. Technol.* 43 (2008) 362-375.

[46] Langmuir, The constitution and fundamental properties of solids and liquids, *J. Am. Chem. Soc.* 38 (1916) 2221-2295.

[47] G. Mckay, H. S. Blair, J. R. Gardener, Adsorption of dyes on Chitin-i: Equilibrium studies, *J. Appl. Polym. Sci.* 27 (1982) 3043-3057.

[48] M. M. Rahman, S. B. Khan, H. M Marwani, A. M. Asiri, K. A. Alamry. Selective Iron(III) ion uptake using CuO-TiO2 nanostructure by inductively coupled plasma-optical emission spectrometry, C*hem. Central J.* 6 (2012) 158-163.

[49] S. B. Khan, M. M. Rahman, H. M. Marwani, A. M. Asiri, K. A. Alamry, M. A. Rub. Selective Adsorption and Determination of Iron(III): Mn_3O_4/TiO_2 Composite Nanosheets as Marker of Iron for Environmental Applications, *App. Surf. Sci.* 282 (2013) 46-51.

In: ZnO Nanostructures
Editor: Dana Crawford

ISBN: 978-1-53616-773-3

Chapter 3

ELECTROCHEMICAL BIOSENSORS BASED ON NANOSTRUCTURED ZNO FOR PATHOGENS DETECTION

Matteo Tonezzer[1,*] and Dang Thi Thanh Le[2,*]
[1]IMEM-CNR, sezione di Trento, Povo (TN), Italy
[2]ITIMS, Hanoi University of Science and Technology (HUST), Hanoi, Vietnam

ABSTRACT

Pathogens, infectious microbes that spread diseases, are one of the leading causes of death among youngs, causing several million deaths per year. Pathogens detection in human environment (especially indoor) is therefore crucial to reduce the risk and save human lives. Furthermore, small, cheap and portable devices, which can be used by personnel with little or no training, are crucial for early diagnosis in areas without hospital or points-of-care. Electrochemical sensors are ideal candidates for this role, and ZnO nanostructures are proving very useful for optimizing their performance. In this chapter we will briefly describe the structure of this

* Corresponding Author's Email: matteo.tonezzer@cnr.it, thanhle@itims.edu.vn.

type of biosensors, and then show the different types of ZnO nanostructures that are used in this field, grown through different methods. We will list scientific papers focusing on the detection of different pathogens and analytes, reporting and comparing the performance of the sensors prepared by various groups around the world. The use of nanoparticles, nanorods, nanowires, nanotubes, nanofibres, nanotetrapods and other morphologies at the nanometer level will be described. The electrochemical biosensors discussed in this chapter detect the concentration of glucose inside cells, presence of hepatitis B virus, E. coli, H1N1 swine influenza virus, cholesterol, Legionella pneumophila, uric acid, C-reactive proteins, N. meningitidis, different Leptospira species, urea, avian influence, and others. We think that this overview of electrochemical biosensors using ZnO nanostructures in one of their components can be useful for approaching this field.

Pathogens are infectious microbes that can cause disease, especially in young people. Even today, pathogens are the leading cause of death among children and young adults, with about 13 million deaths a year [1]. The most common pathogens are bacteria, viruses and fungi. Their presence is dangerous to health even in small quantities. For this reason, a biosensor must be specific (detect the specific pathogen) but also very sensitive, able to “pick the needle out of a haystack” [2]. International Union of Pure and Applied Chemistry (IUPAC) defines biosensor as a “device that uses specific biochemical reactions mediated by isolated enzymes, immunosystems, tissues, organelles or whole cells to detect chemical compounds usually by electrical, thermal or optical signals” [3]. For this reason, an electrochemical sensor is the ideal base for a biosensor. Electrochemical techniques embrace a wide range of analytical options, today enhanced by new technological improvements in signal processing and front-end electronic systems.

Generally, electrochemistry implies the transfer of charge between an electrode and a liquid. Both the reactions at the interface and the electrical conduction inside the bulk solution contribute to this process. In electrochemistry, the reaction under investigation can generate a measurable current (in which case it is called amperometry), a measurable potential or charge accumulation (named potentiometry) or a measurable alteration of the conductivity properties of a medium (conductometry), between electrodes [4]. Electrochemical sensing usually involves a *reference* electrode, a *counter* (or auxiliary) electrode and a *working* electrode, known as the sensing or redox electrode as well. A number of different electrochemical techniques can be used for biosensing applications, such as potentiometric (quantifying variations in open circuit potential), amperometric (measuring currents given by the reduction or oxidation of electroactive species) and impedimetric sensors (determining the impedance of the system upon immobilization of biomolecules at the electrode surface) [5]. Typically, if the current is measured at a constant potential, this is referred to as *amperometry*, while when the current is measured during controlled variations of the potential, this is referred to as *voltammetry* [4]. Potentiometric devices, instead, measure the accumulation of a charge potential at the working electrode compared to the reference electrode in an electrochemical cell when zero or negligible current flows between them.

Biosensors aim to exploit the power of electrochemical techniques to evaluate biological processes by quantitatively producing an electrical signal that relates to the concentration of a specific biological analyte. Examples of recent applications of impedance measurement techniques regard the usage of faradaic impedance spectroscopy for developing enzyme sensors, immunosensors and deoxyribonucleic acid (DNA) sensors: the faradaic impedance spectroscopy has been proposed in association with chrono-potentiometry as a mean to follow biocatalytic processes.

Electrochemical biosensors are probably the most important and common type of biosensor. The other types are optical, calorimetric, piezoelectric and acoustic. Electrochemical biosensors respond to electron transfer, electron consumption, or electron generation during a biochemical interaction. This class of sensors is of key importance as it is simpler to miniaturize than most other biosensors, making them ideal for point-of-care applications.

Typically, the structure of a biosensor includes a substrate (silicon, glass or a polymer), a conductive layer (polysilicon, silica, silicon nitride, a metal or a metal oxide) and above all a layer of molecules for specific capture (antibodies, enzymes, DNA/RNA probes, phage-derived biomolecular recognition probes…). This last layer is fundamental, since the specificity of the biosensor is given by the particular molecules that recognize antigens or epitopes on the exterior of the target pathogen. The recognition elements are typically bound to the surface of the nanomaterials and their interaction with pathogens is monitored via a signal transduction mechanism.

As can be seen in Figure 1, both the electrochemical sensors and the biosensors have received more and more interest in recent years, proving to be interesting for the detection of pathogenic bacteria. Figure 1 shows the number of scientific articles published from year to year and indexed by Scopus and Web of Science.

One of the reasons for this growing interest is the increase in their performance, which is certainly due also to the use of nanostructured materials, which greatly increase the active surface of the device.

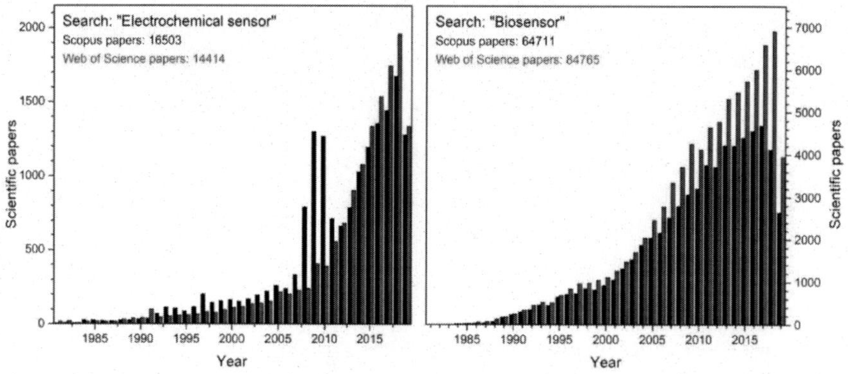

Figure 1. Number of scientific articles published over the years. The two researches "electrochemical sensor" and "biosensor" were carried out in Scopus and Web of Science. It should be noted that the year 2019 includes only the articles until the 24th of July.

A variety of nanostructured metal oxides (NMOs) are currently used in biosensor applications and attracting more and more interest. Nanomaterials are appealing for every type of biosensor, based on different transduction mechanism [2]. The choice of nanomaterial and architecture to use in a specific biosensor is typically dictated by the signal transduction mechanism one wants to employ.

For the fabrication of a high-performance biosensor, the role of an immobilization matrix is very important. Among the various possible matrices, metal oxide nanostructures are attracting interest due to their excellent optical and electrical properties due to electron and phonon confinement, high surface-to-volume ratio, different surface stoichiometry, intense surface reaction activity, robust catalytic efficiency and strong adsorption ability [1, 6, 7]. Nanomaterials with engineered morphology and size, tuned functionalities like adsorption capability and high biomolecule loading capacity, demonstrated to achieve enhanced electron transport between the biomolecule and the electrode. The bio-friendly microenvironment provided by a nanostructured material can help a biomolecule to maintain its conformation and maximum biological activity, giving the biosensor enhanced signal transduction and stability. These distinctive properties of nanomaterials could be exploited for interfacing biological recognition with electronic signal transduction for designing a new generation of bioelectronics devices.

Biosensors are characterized by interesting features: real-time, on-site and simultaneous detection of multiple pathogens utilizing different selective biomolecules and the processing power of modern electronics. Based on the transduction mechanism, biosensors can be classified as electrochemical, optical, piezoelectric, and thermal. Among the various types of biosensors, electrochemical sensors have recently attracted substantial interest as bioanalytical devices, playing an important role in the detection of specific pathogens [3]. Electrochemical sensors are based on measuring variations of an electric signal (current, potential or impedance) induced by biochemical reactions.

ZnO is a well-known n-type II–VI semiconductor, widely used in many transversal applications thanks to its excellent properties: direct band gap of 3.37 eV, high electron mobility (210 $cm^2 \cdot V^{-1} \cdot s^{-1}$) and a large exciton binding energy of 60 meV at room temperature [8, 9].

As a transducer for biosensors, ZnO is advantageous for a number of reasons: 1) is it an inexpensive and ecological material with very low toxicity to human body, 2) it can be obtained through low-temperature synthesis, 3) it is very easy to be grown in a variety of nano-morphologies, and 4) it is transparency to visible light.

Nanostructured zinc oxide also shows other properties that make him highly promising for applications in electrochemical biosensing: strong adhesion ability, high surface area, strong ion adsorption capability (high isoelectric point, IEP ~ 9.5), good catalytic efficiency (oxygen storage capacity), long chemical stability, high resistance against corrosion and oxidation, potentially very small grain size. As mentioned, the high surface-to-volume ratio of ZnO nanostructures can accommodate big loads of analytes and offers a favorable microenvironment to biomolecules, preserving their activity also once immobilized. Moreover, ZnO nanostructures can fully exploit the electrochemical ability of biomolecules, thanks to their excellent electron transfer rate. Zinc oxide, which is a biomimetic semiconducting material, can be integrated with recent micro-electro-mechanical systems (MEMS) and solid-state technologies for the fruitful miniaturization of biosensors.

ZnO nanostructures with very different morphologies have been synthesized via physical and chemical methods for biosensing applications. These nanoscale structures include nanoparticles (NPs) [10-13], nanorods (NRs) [14-21], nanowires (NWs) [22-28], nanotubes (NTs) [29-33], nanofibers (NFs) [34, 37], nanotetrapods [38-40], nanospheres [41], nanocombs [42], nanoflowers [43-45], nanonails [46], nanoflakes [47], nanocomposites [48-57] and thin films [58].

ZnO-based matrices with different shapes have been used for binding various biosensing molecules such as glucose oxidase (GOx), cholesterol oxidase (ChOx), uricase, tyrosinase, microperoxidase, horseradish peroxidase (HRP), biotin, myoglobin, single-stranded DNA (ssDNA), different antibodies, enzymes, whole cells, antigens and other proteins for the detection of their respective analytes. ZnO based biosensors have been developed with various architectures, like field effect transistors (FETs) [25, 22, 19, 59], ion-sensitive field effect transistors (ISFETs) [59, 60], optical and piezoelectric devices (SAW, QCM, …) and electrochemical transducers [61].

NANOPARTICLES

In scientific literature, the most used techniques to grow ZnO nanoparticles for biosensing applications, are wet chemical, hydrothermal and solvothermal methods. Nanosized particles of ZnO realized by wet chemical methods can increase the efficiency of photochemical reactions and greatly enhance the catalytic activity of glucose oxidase, making possible to produce novel photoelectrochemical systems [10]. In the work of Ren and collaborators, when glucose concentration is 10 mM, the presence of ZnO nanoparticles in the GOx electrode increases the current from 0.82 to 21 $\mu A \cdot cm^{-2}$. Furthermore, UV irradiation of the enzyme electrode for 2 h, increases the current response of about 30% and lowers the detection limit about 20 times compared to that in the dark (from 1.1×10^{-4} M to 5.6×10^{-6} M). ZnO nanoparticles with a diameter around 30 nm were synthesized via hydrothermal route and added into a polymer system (polyvinyl ferrocenium) to detect the Hepatitis B Virus (HBV) via the electrochemical detection of nucleic acid hybridization [11]. ZnO nanoparticles increased the electrochemical response of the polymer-modified electrode and lowered the detection limit down to 11.7 μg/mL. A simple and cost-effective glucose biosensor was developed for working at low potential and reduce the importance of interferences commonly found in real samples. The ZnO nanoparticles on the carbon paste electrode provided a biocompatible micro-environment for GOx and also a pathway for electron transfer between GOx and the electrode itself [12]. A relatively low detection limit of 9.1×10^{-3} mM was found, a Michaelis–Menten constant K_m of 0.124 mM, while the maximum current was 2 μA. Real human serum samples were also tested with the intention of check the practical utility of this glucose biosensor, succeeding with a confidence of 95%. ZnO nanoparticles with very different morphologies have been synthesized by solvothermal process adding various amino acids (lysine, arginine and cysteine) to a water-ethanol solution (1:1 in volume) zinc acetate [13]. Nine diverse resulting ZnO nanostructures were used to decorate multi-walled carbon nanotubes (MWCNT) on glassy carbon electrode (GCE), and the hybrid nanomaterial was used as new nanobiosensors for the detection of glucose without using any GOx or nafion. Spherical ZnO nanoparticles showed the highest sensitivity (64.29 $\mu A \cdot cm^{-2} \cdot mM^{-1}$), good repeatability and selectivity (no interference from dopamine, uric acid and fructose).

NANORODS

Recently, several publications focused on ZnO nanorods and nanowires for biosensing applications. Most of them use hydrothermal method to synthesize the ZnO nanosctructures. Asif and collaborators developed a biosensor based on hexagonal ZnO nanorods grown on the tip of a Ag-covered glass capillary for intracellular measurement of glucose in single human adipocytes and frog oocytes [14]. The proposed intracellular biosensor showed a quick response time (less than 1 s) and a wide linear range (from 0.5 to 1000 μM) with sensitivity of 42.5 mV/decade at 23°C. Ferrocene-functionalized ZnO nanorods were synthesized by hydrothermal method for a novel amplified electrochemical immunoassay to detect *E. coli* [15]. The nanorods were employed for immobilization of $_{d}$Ab and Fc with a typical Stöber-based surface modification, obtaining {$_{d}$Ab–ZnO–Fc} bioconjugates. The immunoassay showed an excellent detection of *E. coli* with a wide linear range (10^2 - 10^6 cfu/mL) and a limit of detection of 50 cfu/mL. Exploiting an initial pre-enrichment step, the technique allowed to detect down to 5 cfu/10 mL *E. coli* in hospital sewage water. ZnO nanorods grown in aqueous phase have been used for detecting urea with quasi-reversible behavior and optimum activity at pH 7.0 and room temperature (20-30°C) [16]. The sensor showed a linear behavior in the range 1–20 mM of urea, a high sensitivity (0.4 $\mu A \cdot mM^{-1}$), a response time of 3 seconds, a detection limit of 0.13 mM, and good selectivity to urea even in presence of glucose and ascorbic acid.

A 3D patterned ZnO nanorod network was used as electrode for an immunosensor detecting H1N1 swine influenza virus [17]. The sensor usesd sandwich enzyme-linked immunosorbent assay (ELISA) method, and was tested through cyclic voltammetry. The porous structure of the ZnO nanorods network provides a large surface area for proteins binding, higher entrapment, and long-term stability. The reduction current peak at 0.25 V increased logarithmically from 259.37 to 577.98 nA as the H1N1 SIV concentration raised from 1 pg/mL to 5 ng/mL. The shape of zinc oxide nanostructures is very important in several application, including sensing and biosensing. In order to investigate this topic, Ahmad et al. fabricated and investigated the performance of glucose biosensors based on ZnO nanorods with different aspect ratios [18]. The biosensor based on the nanorods with the largest aspect ratio (60) demonstrated the highest sensitivity (110.76 $\mu A/mM\ cm^2$), with very fast response time (less than 1 s). The fabricated biosensor also showed promising results for selective detection of glucose in urine, which is useful for spotting renal glycosuria. The same group used, for the first time, vertically aligned ZnO nanorods for the fabrication of a solution-gated field-effect-transistor (FET) sensor for real-time detection of cholesterol [19]. The fabricated sensor showed a high sensitivity of 10 $\mu A \cdot cm^{-2} \cdot mM^{-1}$ and a detection limit of 0.05 μM. The sensor showed low interference from electroactive species (AA, UA, glucose, L-Cyst) and could detect cholesterol in human serum and whole blood. A matrix of ZnO nanorods was the basis of an ultrasensitive electrochemical immunosensor developed by Park and coworkers for detecting *Legionella pneumophila* [20]. An ELISA was performed using the fabricated immunosensor, evaluated then by cyclic voltammetry in 1-5000 pg/mL *L. pneumophila* concentration. The results indicated that the high isoelectric point (IEP) of ZnO NRs (~ 9.5) offers a positively charged surface that can facilitate the immobilization of low IEP (~5) PAL protein (an immune-dominant component of *Legionella* antigens), while preserving the antibody and the conjugated HRP enzyme activity. The calculated limit of detection was ~ 1 pg/mL, lower than the previously reported 5 pg/mL limit from carbon nanotubes-modified electrodes, with a signal to noise ratio higher than 4. The excellent selectivity (negligible response to bovine serum albumin BSA, H1N1 and Thrombin) was due to the highly specific binding affinity of the immunoreactions between antigen and antibody, as well as the minimization of non-specific adsorption by the efficient blocking.

An electrochemical immunosensor using microfluidic was improved by growing ZnO nanorods on the inner surface of the PDMS chamber [21]. Splitting the flow into 3 channels, the sensor was able to detect H1N1, H5N1, and H7N9 influenza viruses simultaneously. The sensor was able to selectively detect each antigen, showing a linear range of 1 pg/mL ÷ 10 ng/mL and a low limit of detection of 1 pg/mL.

NANOWIRES

A field effect transistor based on ZnO nanowires was used as biosensor for the detection of biotin-streptavidin biomolecular interactions [22]. The presence of streptavidin with concentrations from 0 to 250 nM lead to electrical current changes of up to 22.5 nA. Even at low concentration (2.5 nM), the ZnO NW biosensor could detect the streptavidin with a current increase of 7.5 nA, meaning that the detection limit is lower than the nM range, making the biodevice ideal for sensitive, real time, and label-free detection of biological species. A plastic substrate coated with gold was modified with ZnO nanowires and uricase to obtain an uric acid biosensor [23]. The aligned nanowires were used as a surface where to immobilize uricase electrostatically. The biosensor showed high sensitivity (32 mV/decade), a linear measurement range of 1 – 1000 μM and a quick response (less than 10 s). A urea biosensor with high sensitivity was realized growing well-aligned ZnO nanowire arrays on gold-coated plastic substrates. The zinc oxide nanostructures offered a suitable microenvironment for enzymes, that preserve their activity because of the strong electrostatic interaction between ZnO and urease [24]. The sensor showed a sensitivity of 52.8 mV/decade for 0.1 ÷ 40 mM urea, a detection limit of 0.1 mM and a fast response time (less than 4 s). The biodevice also showed a good selectivity with only minor response to common interferents (ascorbic acid, uric acid, glucose, potassium and sodium ions), good reproducibility and stability over time and reuses. A single ZnO nanowire was used as a biologically sensitive field-effect-transistor (BioFET) for the detection of uric acid on a wide range of concentrations [25]. The single-nanowire bioFET sensor could easily detect uric acid down to a concentration of 1 pM with a conductance increase of 14.7 nS and a response time in the order of milliseconds.

Zhao and coworkers designed and fabricated an electrochemical biosensor for the detection of uric acid based on an individual ZnO micro/nanowire [26]. They considered that the performance of a biosensor based on numerous nanostructures could be endangered by the electrical discontinuity resulting from the many resistive contacts between them. A biosensor based on one single wire, on the other hand, exhibits high sensitivity due to the excellent electron transport properties of the crystalline nanostructure to the electrode, while exploiting the excellent biocompatibility of ZnO. The realized biosensor displays a high sensitivity of 89.74 $\mu Acm^{-2} \cdot mM^{-1}$, a linear behavior in the range 0.1 – 0.6 mM and a limit of detection of 25.6 μM for uric acid detection. Zinc oxide nanowires were grown by chemical vapor deposition and functionalized with lactate oxidase to exploit their large surface to volume ratio, biocompatibility and high isoelectric point [27]. The obtained L-lactic acid sensor displayed a good sensitivity (15.6 $\mu A/cm^2 \cdot mM$), a linear range of 12 μM – 1.2 mM, a limit of detection of 12 μM and a good specificity (negligible response to glucose, uric acid and ascorbic acid). ZnO nanowires were synthesized directly on a paper substrate in order to build an electrochemical microfluidic sensing device for glucose detection in human serum [28]. The semiconducting nanostructures were used to decorate the working electrode, providing high surface-to-volume ratio and high enzyme-capturing efficiency. The sensor based on ZnO NWs showed a sensitivity of 2.88 $\mu A \cdot cm^{-2} \cdot mM^{-1}$ and a limit of detection of 94.7 μM in phosphate-buffered saline (PBS).

NANOTUBES

As mentioned, size and morphology of ZnO nanostructures is important to achieve better sensing performance. This is proven, for example, by the comparison of amperometric response of hexagonal ZnO nanorods and nanotubes (NTs) with similar diameters [29]. A glucose biosensor was fabricated via immobilization of GOx onto ZnO nanotube arrays through cross-linking method. The ZnO nanotubes were obtained by chemical etching of hexagonal ZnO nanorods that had been electrochemically deposited on a gold surface. The glucose biosensor showed a fast response time (3 s), a wide linear range (from 50 μM to 12 mM), good sensitivity of 21.7 $\mu A/mM \cdot cm^2$ and a limit of detection of 1 μM.

Arrays of highly oriented single-crystal ZnO nanotubes were also grown on glass coated with indium-doped tin oxide (ITO), in order to detect glucose [30]. The nanotubes were biofunctionalized with glucose oxidase in conjunction with a Nafion coating, and used as a working electrode for an enzyme-based glucose biosensor. The biosensor exhibited high sensitivity (30.85 $\mu A/mM \cdot cm^2$), wide linear range (from 10 μM to 4.2 mM), and a low limit of detection (10 μM). Also Ali and his group developed a glucose biosensor based on highly oriented single-crystal ZnO-NT arrays [31]. The nanotubes were realized on a gold-coated glass substrate, and had a diameter of 100-20 nm and a length around 1 μm. The electrochemical response of the sensor was linear over a quite wide logarithmic concentration range (from 0.5×10^{-6} to 1.2×10^{-2} M) with high sensitivity (69.12 mV/decade). Ibupoto et al. developed a potentiometric biosensor exploiting antibodies immobilized on ZnO nanotubes for the detection of C-reactive protein (CRP) [32]. The anti-C-reactive-protein was immobilized on the ZnO NTs, providing the capturing environment for CRP in the formation of complex between antibody and CPR. The potentiometric biosensor was able to detect CRP in the concentration range of 1.0×10^{-5} - 1.0×10^{0} mg/L, with a sensitivity of 13.17 ± 0.42 mV/decade. The piezoelectricity of ZnO helped the formation of complex between immobilized antibodies and CRP on the surface of the nanotubes. Multiwall ZnO nanotubes, synthesized by replicating silk micro/nanostructure, are an interesting platform to immobilize GOx and detect glucose [33]. The mesopores in the walls of silk-like ZnO NTs are larger than GOx (20-50 nm compared to 12-20 nm), and are able to accept high quantities of enzymes without losing activity. The sensor exhibits high sensitivity (47.2 $\mu A/mM \cdot cm^2$), low limit of detection (10 μM), fast response time (< 2s) and quite low K_m (1.09 mM).

NANOFIBERS

Also polycrystalline nanostructures, as nanofibers, are used in this field. One single ZnO nanofiber was used by Ahmad and collaborators [34] to produce an amperometric glucose biosensor with high sensitivity. The biosensor showed good and reproducible performance, with a sensitivity of 70.2 μA/mM·cm^2, low limit of detection (1 μM) and short response time (< 4 s). Furthermore, it also exhibits good selectivity (negligible response to cholesterol, AA, L-Cys and urea) and long-term stability (more than 4 months). A platinum electrode was modified with ZnO electrospun fibers and phosphotungstic acid, in order to detect dopamine in presence of ascorbic acid [35]. The fiber underwent an annealing at 400°C in air before use. The device showed a linear response along the range 1.9×10^{-7} - 4.5×10^{-4} M, a limit of detection of 8.9×10^{-5} M and a good stability (a peak decrease of 12.5% after one month). Zinc oxide nanofibers were also modified with a self assembled monolayer (SAM) of 3-mercaptopropionic acid, with the aim of providing a high density of carboxylic acid groups, to be functionalized with protein molecules via simple crosslinking chemistry [36]. The sensor was tested with standard biotin-streptavidin interaction as a proof of concept, showing good performance: good sensitivity (613 μA/mg/mL/cm^2), wide linear range (1 μg/mL – 1 fg/mL) and a low limit of detection (1 fg/mL). Paul et al. demonstrated an ultrasensitive malaria electrochemical immunoassay biosensor platform utilizing Cu-doped ZnO electrospun nanofibers functionalized with Mercaptopropylphosphonic acid (MPA) [37]. The impedimetric detection response of the electrode modified with Cu-doped ZnO nanofibers shows excellent sensitivity of 28.5 KΩ/(gm/ml)/cm^2 in the detection range of 10 ag/ml - 10 μg/ml, and a detection limit of 6 attogram/ml.

NANOTETRAPODS

ZnO nanotetrapods (sort of tiny caltrops) were used to produce a glucose biosensor, exploiting their good electron transduction and positively charged surface, where negatively charged GOx was immobilized via electrostatic interaction, without any additional mediator [38]. The biodevice exhibits high sensitivity of (25.3 μA/mM·cm^2), low detection limit (4 μM), linear response from 5μM to 6.5 mM and high stability, which is attributed to their multi-terminal electron communication feature. ZnO tetratpods were also grown via vapor phase transport method in order to exploit their high surface-to-volume ration and their multi-terminal and detect L-lactic acid [39]. The simple and low-cost biosensor was obtained functionalizing the tetrapods with lactate oxidase and nafion. It presents linear response from 3.6 μM to 0.6 mM, high sensitivity of 28.0 μA·cm^{-2}·mM^{-1}, detection limit of 1.2 μM, and low Km of 0.58 mM. *Helicobacter pylori* virulence factor CagA (cytotoxin-associated gene A) was detected with a biohybrid interface consisting on screen printed electrodes coated with ZnO nanotetrapods on which CagA antigen was immobilized. The nanotetrapods were ion irradiated with N_2^+ ions in order to improve their electroconductivity, and the sensor sensitivity and specificity [40]. The biosensor showed good linearity (0.2 to 50 ng/mL) and low limit of detection (0.2 ng/mL).

OTHER NANOSHAPES

ZnO hollow nanosphere prepared by hydrothermal method using carbon spheres as template proved to be a very good platform to immobilize glucose oxidase for glucose biosensor [41]. This morphology gave better results (higher catalytic activity) than the same setup using nanoflowers, demonstrating that shape is important. The hollow nanospheres showed linear response over a concentration range from 5.0 μM to 13.15 mM, a detection limit of 1.0 μM, and a sensitivity of 65.82 μA/(mMcm^2).

Table 1. Performance of biosensors based on ZnO nanostructures, listed by method and morphology

Detection method	ZnO nanostructure	Immobilized biomolecule	Detected analyte	Linear range and LoD	Response time	Reference (Year)
Amperometry	Nanoparticles	GOx	Glucose	5.6×10^{-6} M (under illumination)		[10] (2009)
	Nanoparticles	ssDNA	HBV DNA	20 ÷ 140 μg/mL 11.7 μg/mL	-	[11] (2011)
	Nanoparticles	Glucose oxidase Non-enzyme	Glucose	9.1×10^{-3} ÷ 14.5 mM 9.1×10^{-3} mM 1 ÷ 10 mM 0.82 mM	20 s -	[12] (2013) [13] (2015)
	Nanorods	Antibody Ab	E.coli	10^2 ÷ 10^6 cfu/mL 50 cfu/mL	-	[15] (2011)
	Nanorods	Urease	Urea	1 ÷ 20 mM 0.13 mM	3 s	[16] (2011)
	Nanorods	H1N1 antibody	H1N1 SIV	1 pg/mL ÷ 5 ng/mL 1 pg/mL	-	[17] (2012)
	Nanorods	GOx	Glucose	0.3 μM	< 1 s	[18] (2012)
	Nanorods	Legionella antibody	Legionella pneumophila	1 pg/mL ÷ 5 ng/mL 1 pg/mL	-	[20] (2014)
	Nanorods	 H1N1 antibody H5N1 antibody H7N9 antibody	 H1N1 H5N1 H7N9	1 pg/ml ÷ 10 ng/ml 1 pg/ml 1 pg/ml 1 pg/ml		[21] (2016)
	Nanowires	Uricase Cy5 labeled uricase	Uric acid Uric acid	1 pM ÷ 0.5 mM 1 pM 0.1 ÷ 0.59 mM 25.6 μM	~ millisecond -	[25] (2012) [26] (2013)
	Nanowires	Lactate oxidase	L-lactic acid	12 μM ÷ 1.2 mM 12 μM	-	[27] (2014)
	Nanowires	Glucose oxidase	Glucose	0 ÷ 15 mM 94.7 μM	-	[28] (2015)
	Nanotubes	Glucose oxidase	Glucose	50 μM ÷ 12 mM 1 μM	3 s	[29] (2009)

Table 1. (Continued)

Detection method	ZnO nanostructure	Immobilized biomolecule	Detected analyte	Linear range and LoD	Response time	Reference (Year)
				10 μM ÷ 4.2 mM 10 μM 0.02 ÷ 4.4 mM 10 μM	< 6 s < 2 s	[30] (2009) [33] (2013)
	Nanofibers	GOx	Glucose	0.25 ÷ 19 mM 1 μM	< 4 s	[34] (2010)
	Nanofibers	Phosphotungstic acid	Dopamine	$1.9 \times 10^{-7} \div 4.5 \times 10^{-4}$ M 0.089 μM	-	[35] (2013)
	Nanofibers	Biotin	Streptavidin	1 μg/mL ÷ 1 fg/mL 1 fg/mL	-	[36] (2015)
	Nanofibers	Histidine-rich protein-2 Antibody	Plasmodium falciparum	10 ag/mL ÷ 10 μg/mL 6.8 ag/mL	-	[37] (2016)
	Nanotetrapods	Glucose oxidase	Glucose	0.005 ÷ 6.5 mM 4 μM	< 6 s	[38] (2010)
	Nanotetrapods	Lactate oxidase	L-lactic acid	3.6 mM ÷ 0.6 mM 1.2 mM	10 s	[39] (2012)
	Nanospheres	GOx	Glucose	0.005 ÷ 13.15 mM 1 μM	< 5 s	[41] (2011)
	Nanocombs	GOx	Glucose	0.02 ÷ 4.5 mM 0.02 mM	< 10 s	[42] (2006)
	Nanoflowers	ssDNA	DNA (meningitidis)	5 ÷ 240 $ng \cdot \mu l^{-1}$ 5 ng/μl	35 s	[43] (2014)
	Nanoflowers	Urease	Urea	1.65 ÷ 16.50 mM 0.9 mM	4 s	[45] (2015)
	Nanonails	GOx	Glucose	0.1 ÷ 7.1 mM 5 μM	< 10 s	[46] (2008)
	Nanoflakes	GOx	Glucose	500 nM ÷ 10 mM	< 4 s	[47] (2010)
	Nanoclusters	GOx	Glucose	20 μM	-	[48] (2007)

Detection method	ZnO nanostructure	Immobilized biomolecule	Detected analyte	Linear range and LoD	Response time	Reference (Year)
	Nanocomposites	Cholesterol oxidase	Cholesterol	5 ÷ 300 mg.dL^{-1} 5 mg.dL^{-1}	15 s	[49] (2008)
		DNA	H1N1	-	-	[54] (2016)
		GOx	Glucose	0.1 ÷ 33 μM 10 nM 1 ÷ 15 mM 0.04 mM 0.5 ÷ 8 mM 2.5 μM 1 ÷ 25 mM 12 μM 0.05 ÷ 0.3 mM	< 5 s - < 5 s < 5 s -	[50] (2010) [51] (2012) [53] (2012) [55] (2016) [56] (2016)
Potentiometry	Nanorods	GOx	Glucose	0.5 ÷ 1000 μM	< 1 s	[14] (2010)
	Nanowires	Uricase	Uric acid	1 ÷ 100 μM	6-9 s	[23] (2011)
	Nanowires	Urease	Urea	0.1 ÷ 40 mM 0.1 mM	< 4 s	[24] (2011)
	Nanotubes	Glucose oxidase	Glucose	0.5 μM ÷ 12 mM	< 4 s	[31] (2011)
	Nanotubes	C-reactive antibody	C-reactive protein	$1.0 \times 10^{-5} \div 1.0 \times 100$ mg/L 13.17 ± 0.42 mV/decade	-	[32] (2012)
	Nanoflakes	GOx	Glucose	500 nM ÷ 10 mM	< 4 s	[47] (2010)
	Nanocomposites	GOx	Glucose	2 μM ÷ 1.2 mM 0.2 μM	3 s	[52] (2012)
Conductometry	Nanowires	Uricase	Uric acid	1 pM ÷ 0.5 mM 1 pM	~ millisecond	[14] (2012)
	Nanorods	Cholesterol oxidase	Cholesterol	0.001÷ 45 mM 0.05 μM	-	[19] (2013)
	Nanowires	Biotin	Streptavidin	0 ÷ 250 nM 2.5 nM	-	[22] (2010)
	Nanocomposites	ssDNA	Dengue virus	1×10^{-6} M ÷ 100×10^{-6} M 4.3×10^{-5} M	-	[57] (2017)
Impedimetry	Nanoparticles	ssDNA	HBV DNA	20 ÷ 140 μg/mL		[11] (2011)

Table 1. (Continued)

Detection method	ZnO nanostructure	Immobilized biomolecule	Detected analyte	Linear range and LoD	Response time	Reference (Year)
	Nanofibers	Histidine-rich protein-2 Antibody	Plasmodium falciparum	10 ag/mL ÷ 10 μg/mL 6.8 ag/mL	-	[37] (2016)
	Nanotetrapods	Cytotoxin associated gene A (CagA)	Helicobacter pylori	0.2 ÷ 50 ng/mL 0.2 ng/mL	-	[40] (2018)
	Nanoflowers	ssDNA	DNA (meningitides)	$5 \div 240$ ng·μl^{-1}	35 s	[43] (2014)
	Nanoflowers	ssDNA	DNA (Leptospira interrograns)	$10^{-6} \div 10^{-13}$ M	1 min	[44] (2015)
	Nanocomposites	Cholesterol oxidase	Cholesterol	$5 \div 300$ mg.dL^{-1} 5 mg.dL^{-1}	15 s	[49] (2008)
	Nanocomposites	GOx	Glucose	1 ÷ 15 mM 0.04 mM	-	[51] (2012)
	Nanocomposites	GOx	Glucose	0.05 ÷ 0.3 mM	-	[56] (2016)
	Thin film	ssDNA, dsDNA	Neisseria meningitidis	5 ng/μl ÷ 200 ng/μl 5 ng/μl	-	[58] (2017)

Single-crystal ZnO nanocombs grown by vapor phase transport were used as porous network with high specific surface area and good biocompatibility for immobilization of GOx at high densities [42]. The sensor gives a low limit of detection of 0.02 mM, a linear range of measure from 20 μM to 4.5 mM, short response time (< 10 s), low K_m of 2.19 mM and good sensitivity of 15.33 $\mu A.cm^{-2} \cdot mM^{-1}$.

Even ZnO nanostructures with the shape of flowers were grown by hydrothermal method and successfully exploited for efficient detection of N. meningitidis [43]. The realized DNA biosensor uses immobilized single stranded thiolated DNA probe specific to this bacterium, to quantify its complementary target in the range of 5–240 $ng \cdot \mu L^{-1}$ with good linearity, excellent sensitivity of 168.64 $\mu A \cdot ng^{-1} \cdot \mu L \cdot cm^{-2}$ and low detection limit of about 5 $ng \cdot \mu L^{-1}$. Perumal et al. realized a novel structure consisting in gold-decorated nanoflowers for impedance sensing, in order to distinguish pathogenic and non-pathogenic *Leptospira* species [44]. The Au-decorated nanoflowers bioelectrode has good linearity, and a detection limit of 100 fM. The resistance to charge transfer (Rct) of t-DNA (1 nM) was nearly 8.5 times larger than that of single-base-mismatched DNA at the same concentration, demonstrating that the DNA biosensor has excellent sequence specificity, even towards a single base mismatch. Flower-like zinc oxide nanostructures was also synthesized via hydrothermal method by Tak and collaborators [45], and exploited as urea biosensor. The fabricated amperometric biosensor shows a linear sensing response to urea over the concentration range of 1.65 to 16.5 mM, an enhanced sensitivity of 131.98 $\mu A \cdot cm^{-2} \cdot mM^{-1}$, relatively low detection limit of 0.90 mM, low K_m of 0.19 mM and a fast response time of 4 s.

Dense single-crystalline zinc oxide nanonails were grown by thermal evaporation and used as supporting matrix to immobilize glucose oxidase and construct an efficient glucose biosensor [46]. The nanonails-based biosensor showed high sensitivity of 24.613 $\mu A \cdot cm^{-2} \cdot mM^{-1}$, response time shorter than 10 s, low K_m of 14.7 mM, and detection limit of 5 μM.

Also ZnO nanoflakes were grown on the head of an Al-coated glass capillary to be used as an intracellular glucose biosensor [47]. The biosensor showed a quick response (less than 4 s), a sensitivity of 65.2 mV/decade, and a logarithmic linear potential difference over a large range of glucose concentration (from 500 nM to 10 mM). The sensing capability of the sensor was checked by monitoring the rise in the intracellular glucose concentration prompted by insulin addition in adipocytes and frog oocytes.

NANOCOMPOSITES

Nanoclusters of ZnO:Co with very small average size (5 nm) and very high porosity, were grown and employed to realize a novel amperometric glucose biosensor by exploiting the high electrocatalytic activity of ZnO and Co and the high surface/volume ratio of the nanoclusters [48]. The biosensor has a high sensitivity (13.3 μA/mM·cm^2) for glucose detection, low K_m (21 mM) and low detection limit (20 μM). A nanostructured thin film composed of ZnO nanoparticles and chitosan (CHIT) has been used to immobilize cholesterol oxidase (ChOx) via electrostatic interaction, to estimate cholesterol concentration [49]. The ChOx/NanoZnO-CHIT bioelectrode exhibits a linear response on a cholesterol concentration range from 5 to 300 mg·dL^{-1}, a K_m value of 8.63 mg·dL^{-1}, a sensitivity of 1.41×10^{-4} A·mg·dL^{-1}, and a detection limit of 5 mg·dL^{-1}. An electrode based on ZnO/Au hybrid nanocomposites was used to enhance the sensitivity for glucose [50]. The obtained biosensor responded linearly to glucose over a concentration range of 0.1 ÷ 33.0 μM, with a very high sensitivity (1492 μA/mM/cm^2), very low detection limit (10 nM), quick response (5 s), and high affinity to GOx (K_m of 0.41 mM).

In order to facilitate the direct electron transfer (DET) between proteins and electrodes, a ZnO/Cu nanocomposite layer with prickly morphology was fabricated on ITO electrodes via a corrosion method without using any organic reagent. The nanostructured layer was then used for immobilization of the enzyme (GOx in this case) in order to fabricate a glucose biosensor [51]. The nanostructured ZnO/Cu layer allowed the device to reach direct electron transfer, with a good heterogeneous electron transfer rate constant of 0.67 ± 0.06 s^{-1}. The prepared biosensor, working without mediators, showed good sensitivity (97 nA/mM), wide linear range (1 – 15 mM), low detection limit (0.04 mM), and fast response (6 s) for the detection of glucose. A core-shell nanocomposite based on zinc oxide encapsulated in chitosan-graft-poly (vinyl alcohol) was used to realize a potentiometric glucose biosensor [52]. The biosensing device exhibited fast surface-modulated reduction–oxidation reaction with a sensitivity of > 0.04 V/μM, a detection limit of 0.2 μM and a response time of 3 s. Another glucose biosensor was fabricated modifying a Pt electrode with a layer of NiO doped ZnO NRs and then immobilizing GOx on it [53]. The sensor showed good linearity, fast response (< 5 s), high sensitivity (61.78 $\mu A/mM\cdot cm^2$) and low detection limit (2.5 μM). An alternative hybrid nanomaterial (graphene/ZnO) was obtained via a simple exfoliation in liquid phase followed by incorporation of zinc oxide. This approach preserved the quality of graphene, allowing the nanocomposite to show improved electrochemical properties. The nanocomposite was then used to detect hydrogen peroxide and Avian Influenza *H5* gene [54]. The biosensor exhibited very high electrocatalytic activity to the reduction of hydrogen peroxide (H_2O_2) on a linear range of 1-15 mM, sensitivity of 3.26 $\mu A\cdot mM^{-1}$, and a limit detection of 7.4 μM. When used as genosensor, the graphene/ZnO device displayed good sensitivity ($P < 0.05$) and high accuracy detecting the *H5* gene PCR amplicon. This approach could be investigated further in order to develop an early and quick detection tool for the Avian Influenza H5N1 virus. Even heterostructures of different NMOs can be used as a substrate for enzyme immobilization towards a biosensor: ZnO NWs were grown on SiO_2 NWs and used to immobilize glucose oxidase, realizing a glucose enzymatic biosensor [55]. The biodevice is stable, reproducible, showing high sensitivity (129 $\mu A\cdot mM^{-1}$) and low detection limit (12 μM). A novel glucose biosensor was realized by imitating biomineralization and growing chitosan-ZnO mesocrystals with different morphologies [56]. Among the different morphologies, starfish-like nanomaterial (with a size of 20-30 nm) showed the best sensitivity (314.0 $\mu A/mM\cdot cm^2$) and good linearity over a range of concentrations from 0.05 to 0.3 μM. A fluorine doped tin oxide glass was modified with a ZnO/Pt-Pd nanocomposite, on which probe DNA was immobilized in order to detect DNA consensus sequence of Dengue virus [57]. The genosensor demonstrated a linear response over a range of 1 to 100×10^{-6} M with a low limit of detection of 4.3×10^{-5} M and a limit of quantification of 9.5×10^{-5} M.

THIN FILMS

A highly porous thin film of nickel doped zinc oxide was used to immobilize single stranded DNA and produce a stable biosensor for the detection of Meningitis [58]. The ssDNA/Ni-ZnO biosensor demonstrated high sensitivity (49.95 μA/decade), a wide linear detection range (5 to 200 ng/μl) and very low limit of detection (5 ng/μl).

As we have shown, zinc oxide nanostructures, with different shapes and size, are an ideal building block for electrochemical biosensors. Basically, the good biocompatibility of ZnO and its rich range of morphology are highly suitable as intermediate layer between electrodes and bio-recognition parts. The huge surface/volume ratio can help the liquid-solid contact, and its high charge carriers mobility makes sure that it is not the bottleneck in the transport of electrons.

ACKNOWLEDGMENTS

This work was supported by the Vietnam National Foundation for Science and Technology Development (NAFOSTED) under grant number 103.02-2015.43.

REFERENCES

[1] Singh, R; Das Mukherjee, M; Sumana, G; Gupta, RK; Sood, S; Malhotra, BD. Biosensors for pathogen detection: A smart approach

towards clinical diagnosis, *Sensors Actuators, B Chem.*, 197, (2014), 385–404. doi:10.1016/j.snb.2014.03.005.

[2] Vikesland, PJ; Wigginton, KR. Nanomaterial enabled biosensors for pathogen monitoring - A review, *Environ. Sci. Technol.*, 44, (2010), 3656–3669. doi:10.1021/es903704z.

[3] Nayak, M; Kotian, A; Marathe, S; Chakravortty, D. Detection of microorganisms using biosensors - A smarter way towards detection techniques, *Biosens. Bioelectron.*, 25, (2009), 661–667. doi:10.1016/j.bios.2009.08.037.

[4] Grieshaber, D; MacKenzie, R; Vörös, J; Reimhult, E. Electrochemical Biosensors - Sensor Principles and Architectures, *Sensors*, 8, (2008), 1400-1458. doi: 10.3390/s80314000.

[5] Hammond, JL; Formisano, N; Estrela, P; Carrara, S; Tkac, J. Electrochemical biosensors and nanobiosensors, *Essays In Biochemistry*, 60, (2016), 69-80. doi: 10.1042/EBC20150008.

[6] Solanki, PR; Kaushik, A; Agrawal, VV; Malhotra, BD. Nanostructured metal oxide-based biosensors, *NPG Asia Mater.*, 3, (2011), 17–24. doi:10.1038/asiamat.2010.137.

[7] Taylor, P; Pandey, P. Prospects of Nanomaterials in Biosensors Prospects of Nanomaterials in Biosensors, *Anal. Lett.*, 41, (2008), 159–209. doi:10.1080/00032710701792620.

[8] Tonezzer, M; Le Dang, TT; Bazzanella, N; Nguyen, VH; Iannotta, S. Comparative gas-sensing performance of 1D and 2D ZnO nanostructures, *Sensors Actuators B Chem.*, 220, (2015), 1152–1160. doi:10.1016/j.snb.2015.06.103.

[9] Le, DTT; Iannotta, S; Hieu, NV; Corradi, C; Huy, TQ; Pola, M; et al., ZnO Nanowires-C Microfiber Hybrid Nanosensor for Liquefied Petroleum Gas Detection, *J. Nanosci. Nanotechnol.*, 14, (2014), 5088–5094. doi:10.1166/jnn.2014.8714.

[10] Ren, X; Chen, D; Meng, X; Tang, F; Hou, X; Han, D; et al., Zinc oxide nanoparticles/glucose oxidase photoelectrochemical system for the fabrication of biosensor., *J. Colloid Interface Sci.*, 334, (2009), 183–7. doi:10.1016/j.jcis.2009.02.043.

[11] Yumak, T; Kuralay, F; Muti, M; Sinag, A; Erdem, A; Abaci, S. Preparation and characterization of zinc oxide nanoparticles and their sensor applications for electrochemical monitoring of nucleic acid hybridization, *Colloids Surfaces B Biointerfaces.*, 86, (2011), 397–403. doi:10.1016/j.colsurfb.2011.04.030.

[12] Aydoğdu, G; Zeybek, DK; Pekyardımcı, Ş; Kılıç, E. A novel amperometric biosensor based on ZnO nanoparticles-modified carbon paste electrode for determination of glucose in human serum, Artif. *Cells, Nanomedicine, Biotechnol.*, 41, (2013), 332–338. doi:10.3109/21691401.2012.744994.

[13] Tarlani, A; Fallah, M; Lotfi, B; Khazraei, A; Golsanamlou, S; Muzart, J; et al., New ZnO nanostructures as non-enzymatic glucose biosensors, Biosens. *Bioelectron.*, 67, (2015), 601–607. doi:10.1016/ j.bios.2014.09.063.

[14] Asif, MH; Ali, SMU; Nur, O; Willander, M; Brännmark, C; Strålfors, P; et al., Functionalised ZnO-nanorod-based selective electrochemical sensor for intracellular glucose, Biosens. *Bioelectron.*, 25, (2010), 2205–2211. doi:10.1016/j.bios.2010.02.025.

[15] Teng, Y; Zhang, X; Fu, Y; Liu, H; Wang, Z; Jin, L; et al., Optimized ferrocene-functionalized ZnO nanorods for signal amplification in electrochemical immunoassay of Escherichia coli, Biosens. *Bioelectron.*, 26, (2011), 4661–4666. doi:10.1016/j.bios.2011.04.017.

[16] Palomera, N; Balaguera, M; Arya, SK; Hernández, S; Tomar, MS; Ramírez-Vick, JE; et al., Zinc Oxide Nanorods Modified Indium Tin Oxide Surface for Amperometric Urea Biosensor, *J. Nanosci. Nanotechnol.*, 11, (2011), 6683–6689. doi:10.1166/jnn.2011.4248.

[17] Jang, Y; Park, J; Pak, YK; Pak, JJ. Immunosensor Based on the ZnO Nanorod Networks for the Detection of H1N1 Swine Influenza Virus, *J. Nanosci. Nanotechnol.*, 12, (2012), 5173–5177. doi:10.1166/ jnn.2012.6361.

[18] Ahmad, R; Tripathy, N; Kim, JH; Hahn, YB. Highly selective wide linear-range detecting glucose biosensors based on aspect-ratio controlled ZnO nanorods directly grown on electrodes, *Sensors*

Actuators B Chem., 174, (2012), 195–201. doi:10.1016/ j.snb.2012. 08.011.

[19] Ahmad, R; Tripathy, N; Hahn, YB. High-performance cholesterol sensor based on the solution-gated field effect transistor fabricated with ZnO nanorods, Biosens. *Bioelectron.*, 45, (2013), 281–286. doi:10.1016/j.bios.2013.01.021.

[20] Park, J; You, X; Jang, Y; Nam, Y; Kim, MJ; Min, NK; et al., ZnO nanorod matrix based electrochemical immunosensors for sensitivity enhanced detection of Legionella pneumophila, *Sensors Actuators B Chem.*, 200, (2014), 173–180. doi:10.1016/j.snb.2014.03.001.

[21] Han, JH; Lee, DY; Chew, CHC; Kim, TH; Pak, JJH. A multi-virus detectable microfluidic electrochemical immunosensor for simultaneous detection of H1N1, H5N1, and H7N9 virus using ZnO nanorods for sensitivity enhancement, *Sens. & Acts B: Chemical*, 228, (2016), 36-42. doi: 10.1016/j.snb.2015.07.068.

[22] Choi, A; Kim, K; Jung, HI; Lee, SY. ZnO nanowire biosensors for detection of biomolecular interactions in enhancement mode, *Sensors Actuators B Chem.*, 148, (2010), 577–582. doi:10.1016/j.snb. 2010.04.049.

[23] Ali, SMU; Alvi, NH; Ibupoto, ZH; Nur, O; Willander, M; Danielsson, B. Selective potentiometric determination of uric acid with uricase immobilized on ZnO nanowires, *Sensors Actuators B Chem.*, 152, (2011), 241–247. doi:10.1016/j.snb.2010.12.015.

[24] Ali, SMU; Ibupoto, ZH; Salman, S; Nur, O; Willander, M; Danielsson, B. Selective determination of urea using urease immobilized on ZnO nanowires, *Sensors Actuators B Chem.*, 160, (2011), 637–643. doi:10.1016/j.snb.2011.08.041.

[25] Liu, X; Lin, P; Yan, X; Kang, Z; Zhao, Y; Lei, Y; et al., Enzyme-coated single ZnO nanowire FET biosensor for detection of uric acid, *Sensors Actuators*, *B Chem.*, 176, (2013), 22–27. doi:10.1016/ j.snb.2012.08.043.

[26] Zhao, Y; Yan, X; Kang, Z; Lin, P; Fang, X; Lei, Y; et al., Highly sensitive uric acid biosensor based on individual zinc oxide

micro/nanowires, *Microchim. Acta.*, 180, (2013), 759–766. doi:10.1007/s00604-013-0981-z.

[27] Zhao, Y; Yan, X; Kang, Z; Fang, X; Zheng, X; Zhao, L; Du, H; Zhang, Y. Zinc oxide nanowires-based electrochemical biosensor for L-lactic acid amperometric detection, *J. Nanopart. Res.*, 16, (2014), 2398. doi: 10.1007/s11051-014-2398-y.

[28] Li, X; Zhao, C; Liu, X. A paper-based microfluidic biosensor integrating zinc oxide nanowires for electrochemical glucose detection, *Microsystems Nanoeng.*, 1, (2015), 15014. doi:10.1038/ micronano.2015.14.

[29] Kong, T; Chen, Y; Ye, Y; Zhang, K; Wang, Z; Wang, X. An amperometric glucose biosensor based on the immobilization of glucose oxidase on the ZnO nanotubes, *Sensors Actuators, B Chem.*, 138, (2009), 344–350. doi:10.1016/j.snb.2009.01.002.

[30] Yang, K; She, G; Wang, H; Ou, X; Zhang, X; Lee, CS; et al., ZnO Nanotube Arrays as Biosensors for Glucose, *J. Phys. Chem. C.*, 113, (2009), 20169–20172. doi:10.1021/jp901894j.

[31] Ali, SMU; Kashif, M; Ibupoto, ZH; Fakhar-e-Alam, M; Hashim, U; Willander, M. Functionalised zinc oxide nanotube arrays as electrochemical sensors for the selective determination of glucose, *Micro Nano Lett.*, 6, (2011), 609. doi:10.1049/mnl.2011.0310.

[32] Ibupoto, ZH; Jamal, N; Khun, K; Willander, M. Development of a disposable potentiometric antibody immobilized ZnO nanotubes based sensor for the detection of C-reactive protein, *Sensors Actuators, B Chem.*, 166-167, (2012), 809–814. doi:10.1016/ j.snb.2012.03.083.

[33] Zhao, M; Li, Z; Han, Z; Wang, K; Zhou, Y; Huang, J; et al., Synthesis of mesoporous multiwall ZnO nanotubes by replicating silk and application for enzymatic biosensor, *Biosens. Bioelectron.*, 49, (2013), 318–322. doi:10.1016/j.bios.2013.05.017.

[34] Ahmad, M; Pan, C; Luo, Z; Zhu, J. A Single ZnO Nanofiber-Based Highly Sensitive Amperometric Glucose Biosensor, *J. Phys. Chem. C.*, 114 (2010) 9308–9313.

[35] Wu, J; Yin, F. Studies on the electrocatalytic oxidation of dopamine at phosphotungstic acid – ZnO spun fiber-modified electrode, *Sensors*

Actuators B. Chem., 185, (2013), 651–657. doi:10.1016/j.snb.2013.05.052.

[36] Paul, B; Prakash, D; Singh, SG; Vanjari, SRK. Highly sensitive SAM modified electrospun zinc oxide nanofiber based label free biosensing platform, *2015 IEEE SENSORS - Proc.*, (2015), 1–4. doi:10.1109/ICSENS.2015.7370418.

[37] Brince Paul, K; Kumar, S; Tripathy, S; Vanjari, SRK; Singh, V; Singh, SG. A highly sensitive self assembled monolayer modified copper doped zinc oxide nanofiber interface for detection of Plasmodium falciparum histidine-rich protein-2: Targeted towards rapid, early diagnosis of malaria, *Biosens. Bioelectron.*, 80, (2016), 39–46. doi:10.1016/j.bios.2016.01.036.

[38] Lei, Y; Yan, X; Luo, N; Song, Y; Zhang, Y. ZnO nanotetrapod network as the adsorption layer for the improvement of glucose detection via multiterminal electron-exchange, Colloids Surfaces A Physicochem. *Eng. Asp.*, 361, (2010), 169–173. doi:10.1016/j.colsurfa.2010.03.029.

[39] Lei, Y; Luo, N; Yan, X; Zhao, Y; Zhang, G; Zhang, Y. A highly sensitive electrochemical biosensor based on zinc oxide nanotetrapods for l-lactic acid detection, *Nanoscale.*, 4, (2012), 3438. doi:10.1039/c2nr30334e.

[40] Chauhan, N; Gupta, S; Avasthi, DK; Adelung, R; Mishra, YK; Jain, U. Zinc Oxide Tetrapods Based Biohybrid Interface for Voltammetric Sensing of Helicobacter pylori, *ACS Appl. Mater. Interfaces.*, 10, (2018), 30631–30639. doi:10.1021/acsami.8b08901.

[41] Fang, B; Zhang, C; Wang, G; Wang, M; Ji, Y. A glucose oxidase immobilization platform for glucose biosensor using ZnO hollow nanospheres, *Sensors Actuators*, *B Chem.*, 155, (2011), 304–310. doi:10.1016/j.snb.2010.12.040.

[42] Wang, JX; Sun, XW; Wei, A; Lei, Y; Cai, XP; Li, CM; et al., Zinc oxide nanocomb biosensor for glucose detection, *Appl. Phys. Lett.*, 88, (2006), 233106. doi:10.1063/1.2210078.

[43] Tak, M; Gupta, V; Tomar, M. Flower-like ZnO nanostructure based electrochemical DNA biosensor for bacterial meningitis detection,

Biosens. *Bioelectron.*, 59, (2014), 200–207. doi:10.1016/j.bios. 2014.03.036.

[44] Perumal, V; Hashim, U; Gopinath, SCB; Haarindraprasad, R; Foo, KL; Balakrishnan, SR; et al., "Spotted Nanoflowers": Gold-seeded Zinc Oxide Nanohybrid for Selective Bio-capture, *Sci. Rep.*, 5, (2015), 12231. doi:10.1038/srep12231.

[45] Tak, M; Gupta, V; Tomar, M. A highly efficient urea detection using flower-like zinc oxide nanostructures, *Mater. Sci. Eng. C.*, 57, (2015), 38–48. doi:10.1016/j.msec.2015.06.052.

[46] Umar, A; Rahman, MM; Kim, SH; Hahn, YB. ZnO Nanonails: Synthesis and Their Application as Glucose Biosensor, *J. Nanosci. Nanotechnol.*, 8, (2008), 3216–3221. doi:10.1166/jnn.2008.116.

[47] Fulati, A; Ali, SMU; Asif, MH; Alvi, NUH; Willander, M; Brännmark, C; et al., An intracellular glucose biosensor based on nanoflake ZnO, *Sensors Actuators B Chem.*, 150, (2010), 673–680. doi:10.1016/ j.snb.2010.08.021.

[48] Zhao, ZW; Chen, XJ; Tay, BK; Chen, JS; Han, ZJ; Khor, KA. A novel amperometric biosensor based on ZnO:Co nanoclusters for biosensing glucose, Biosens. *Bioelectron.*, 23, (2007), 135–139. doi:10.1016/ j.bios.2007.03.014.

[49] Khan, R; Kaushik, A; Solanki, PR; Ansari, AA; Pandey, MK. Malhotra, BD. Zinc oxide nanoparticles-chitosan composite film for cholesterol biosensor, *Anal. Chim. Acta.*, 616, (2008), 207–213. doi:10.1016/j.aca.2008.04.010.

[50] Wei, Y; Li, Y; Liu, X; Xian, Y; Shi, G; Jin, L. ZnO nanorods/Au hybrid nanocomposites for glucose biosensor, Biosens. *Bioelectron.*, 26, (2010), 275–278. doi:10.1016/j.bios.2010.06.006.

[51] Yang, C; Xu, C; Wang, X. ZnO/Cu nanocomposite: A platform for direct electrochemistry of enzymes and biosensing applications, *Langmuir.*, 28, (2012), 4580–4585. doi:10.1021/la2044202.

[52] Shukla, SK; Deshpande, SR; Shukla, SK; Tiwari, A. Fabrication of a tunable glucose biosensor based on zinc oxide/chitosan-graft-poly(vinyl alcohol) core-shell nanocomposite, *Talanta.* 99 (2012) 283–7. doi:10.1016/j.talanta.2012.05.052.

[53] Chu, X; Zhu, X; Dong, Y; Chen, T; Ye, M; Sun, W. An amperometric glucose biosensor based on the immobilization of glucose oxidase on the platinum electrode modified with NiO doped ZnO nanorods, *J. Electroanal. Chem.*, 676, (2012), 20–26. doi:10.1016/j.jelechem.2012.04.009.

[54] Low, SS; Tan, MTT; Loh, HS; Khiew, PS; Chiu, WS. Facile hydrothermal growth graphene/ZnO nanocomposite for development of enhanced biosensor, *Anal. Chim. Acta.*, 903, (2016), 131–141. doi:10.1016/j.aca.2015.11.006.

[55] Miao, F; Lu, X; Tao, B; Li, R; Chu, PK. Glucose oxidase immobilization platform based on ZnO nanowires supported by silicon nanowires for glucose biosensing, *Microelectron. Eng.*, 149, (2016), 153–158. doi:10.1016/j.mee.2015.10.011.

[56] Zhao, S; You, B; Jiang, L. Oriented Assembly of Zinc Oxide Mesocrystal in Chitosan and Applications for Glucose Biosensors, *Cryst. Growth Des.*, (2016), acs.cgd.6b00337. doi:10.1021/acs.cgd.6b00337.

[57] Singhal, C; Pundir, CS; Narang, J. A genosensor for detection of consensus DNA sequence of Dengue virus using ZnO/Pt-Pd nanocomposites, *Biosens. Bioelectron.*, 97, (2017), 75–82. doi:10.1016/j.bios.2017.05.047.

[58] Tak, M; Gupta, V; Tomar, M. An electrochemical DNA biosensor based on Ni doped ZnO thin film for meningitis detection, *J. Electroanal. Chem.*, 792, (2017), 8–14. doi:10.1016/j.jelechem.2017.03.032.

[59] Reyes, PI; Ku, C; Duan, Z; Lu, Y; Solanki, A. ZnO thin film transistor immunosensor with high sensitivity and selectivity, *Appl. Phys. Lett.*, 98, (2011), 173702. doi:10.1063/1.3582555.

[60] Yano, M; Koike, K; Mukai, K; Onaka, T; Hirofuji, Y; Ogata, K; et al., Zinc oxide ion-sensitive field-effect transistors and biosensors, Phys. *Status Solidi.*, 211, (2014), 2098–2104. doi:10.1002/ pssa.201300589.

[61] Flewitt, J; Luo, JK; Fu, YQ; Garcia-Gancedo, L; Du, XY; Lu, JR; et al., ZnO based SAW and FBAR devices for bio-sensing applications, *J.*

Nonnewton. Fluid Mech., 222, (2015), 209–216. doi:10.1016/j.jnnfm.2014.12.002.

Biographical Sketch

Matteo Tonezzer

Affiliation: IMEM-CNR, sede di Trento - FBK, Via alla Cascata 56/C, Povo - Trento, Italy

Education: PhD in Physics

Short biography:

Matteo Tonezzer graduated "cum laude" in Physics of the Matter and received his-PhD degree "with honor" from the Faculty of Physics at the University of Trento, Italy, in 2011. His thesis was the optimization of inorganic and organic nanostructured materials toward gas sensing. In 2011, he won the Young Scientist Award from the European Materials Research Society (EMRS). He worked in research centers in France (ESRF), Brazil (UFMG), Vietnam (HUST), USA (GaTech) and Korea (INHA). He is currently working for IMEM at the Italian National Research Council where is major research interest is synthesis and characterization of nanostructured materials.

Publications from the Last 3 Years:

1. Trinh, Minh Ngoc; Nguyen, Van Duy; Hugo, Nguyen; Matteo, Tonezzer; Chu, Manh Hung; Nguyen, Duc Hoa; Nguyen, Van Hieu*. "*Self-heated Tin Oxide Nanowires Decorated with Silver Catalyst for H_2S Selective Sensor: A Single Sensor with Multisensor Simulated Performance*", published online, doi: 10.1016/j.aca.2019.04.020.

2. Matteo, Tonezzer*. “Selective gas sensor based on one single SnO_2 nanowire”, *Sensors and Actuators B: Chemical*, 288, (2019), 53-59, doi:10.1016/j.snb.2019.02.096.
3. Matteo, Tonezzer*; Jae-Hun, Kim; Jae-Hyoung, Lee; Salvatore, Iannotta; Sang, Sub Kim*. “Predictive gas sensor based on thermal fingerprints from Pt-SnO2 nanowires”, *Sensors and Actuators B: Chemical*, 281, (2019), 670-678, doi:10.1016/j.snb.2018.10.102.
4. Matteo, Tonezzer*; Dang, Thi Thanh Le; Salvatore, Iannotta; Nguyen, Van Hieu. “Selective discrimination of hazardous gases using one single metal oxide resistive sensor”, *Sensors and Actuators B: Chemical*, 277, (2018), 121-128, doi:10.1016/j.snb.2018.08.103.
5. Matteo, Tonezzer*; Dang, Thi Thanh Le; Tran, Quang Huy; Salvatore, Iannotta. “Multiselective visual gas sensor using nickel oxide nanowires as chemiresistor”, *Sensors and Actuators B: Chemical*, 255, (2018), 2785-2793, doi:10.1016/j.snb.2017.09.094.
6. Le*, DTT; Hoang, NV; Hieu, NV; Khue, VQ; Huy, TQ; Tonezzer, M. “Fabrication of electrochemical electrodes based on platinum and ZnO nanofibers for biosensing applications”, *Communications in Physics*, 27, (2017), 221-231, doi: 10.15625/0868-3166/27/3/10517.
7. Matteo, Tonezzer*; Dang, Thi Thanh Le; Tran, Quang Huy; Nguyen, Van Hieu; Salvatore, Iannotta. “Selective hydrogen sensor for liquefied petroleum gas steam reforming fuel cell systems”, *Int. J. Hydrogen Energy*, 42, (2017), 740-748, doi:10.1016/j.ijhydene.2016.11.102.
8. Matteo, Tonezzer*; Dang, Thi Thanh Le; Tran, Quang Huy; Salvatore, Iannotta. “Dual-selective hydrogen and ethanol sensor for steam reforming systems”, *Sensors and Actuators B: Chemical*, 236, (2016), 1011-1019, doi:10.1016/j.snb.2016.04.150.
9. Tonezzer*, M. “Nanomaterials for an Environment 2.0”, *Journal of Nanoscience and Nanotechnology*, 16, (2016), 7849-51, doi: 10.1166/ jnn.2016.12744.

10. Le, DTT; Yu, R; Iacob, E; Tonezzer*, M. "Functionalized ZnO Microbelt as Improved CO Sensor", *Procedia Engineering*, 168, (2016), 1090-1093, doi:10.1016/j.proeng.2016.11.347.
11. Yu, R; Pan, C; Guy, OJ; Dong, L; Tonezzer*, M. "Functional Devices for Clean Energy and Advanced Sensor Applications", *Journal of Nanomaterials*, 2016, (2016), 9162634, doi:10.1155/2016/ 9162634.

Dang Thi Thanh Le

Affiliation: ITIMS, Hanoi University of Science and Technology, Hanoi, Viet Nam

Education: PhD in Materials Science

Short biography:

Dang Thi Thanh Le obtained her MSc and PhD degrees in Materials Science from International Training Institute for Materials Science (ITIMS)-Hanoi University of Science and Technology (HUST), Hanoi, Vietnam in 2001 and 2011. She worked as a visiting postdoc at The Angstrom Laboratory-Uppsala University, Sweden in the academic year of 2011–2012. She is working as a researcher/lecturer at ITIMS. Her current interests include synthesis, characterization and application of nanomaterials for gas-sensing and bio-sensing.

Publications from the Last 3 Years:

1. Pham, Van Tong; Nguyen, Duc Hoa*; Nguyen, Van Duy; Dang, Thi Thanh Le; Nguyen, Van Hieu*. "Enhancement of gas-sensing characteristics of hydrothermally synthesized WO_3 nanorods by surface decoration with Pd nanoparticles", *Sensors and Actuators B*, 223, (2016), 453–460.
2. Nguyen, Van Duy; Nguyen, Duc Hoa*; Nguyen, Thanh Dat; Dang, Thi Thanh Le; Nguyen, Van Hieu*. "Ammonia-gas-sensing characteristics of WO_3/CNT nanocomposites: Effect of CNT

content and sensing mechanism", *Science Advanced Materials*, 8, (2016), 524-533.

3. Nguyen, Van Dung; Dang, Thi Thanh Le*; Nguyen, Dinh Trung; Hoang, Ngoc Dung; Nguyen, Manh Hung; Nguyen, Van Duy; Nguyen, Duc Hoa; Nguyen, Van Hieu**. "CuO Nanofibers Prepared by Electrospinning for Ethanol Gas Sensing Application: Effect of Copper Salt Concentration", *Journal of Nanoscience and Nanotechnology*, 16, (2016), 7910–7918.
4. Matteo, Tonezzer*; Dang, Thi Thanh Le**; Tran, Quang Huy; Salvatore, Iannotta. "Dual-selective hydrogen and ethanol sensor for steam reforming systems", *Sensors and Actuators B*, 236, (2016), 1011–1019.
5. Matteo, Tonezzer*; Dang, Thi Thanh Le**; Tran, Quang Huy; Nguyen, Van Hieu; Salvatore, Iannotta. "Selective hydrogen sensor for liquefied petroleum gas steam reforming fuel cell systems", *International Journal of Hydrogen Energy*, 42, (2017), 740–748.
6. Matteo, Tonezzer*; Dang, Thi Thanh Le; Tran, Quang Huy; Salvatore, Iannotta. "Multiselective visual gas sensor using nickel oxide nanowires as chemiresistor", *Sensors and Actuators B*, 255, (2018), 2785-2793.
7. Nguyen, Van Toan; Chu, Manh Hung; Nguyen, Van Duy; Nguyen, Duc Hoa; Dang, Thi Thanh Le; Nguyen, Van Hieu*. "Bilayer SnO_2-WO_3 nanofilms for enhanced NH_3 gas sensing performance", *Materials Science and Engineering B*, 224, (2017), 163-170.
8. Kien, Nguyen; Chu, Manh Hung; Trinh, Minh Ngoc; Dang, Thi Thanh Le; Nguyen, Duc Hoa; Nguyen, Van Duy; Nguyen, Van Hieu. "Low-temperature prototype hydrogen sensors using Pd-decorated SnO_2 nanowires for exhaled breath applications", *Sensors and Actuators B*, 253, (2017), 156–163.
9. Tran, Thi Mai; Nguyen, Thuy Chinh; Rajesh, Baskaran; Nguyen, Thi Thu Trang; Vu, Viet Thang; Dang, Thi Thanh Le; Do, Quang Minh; Thai, Hoang. "Tensile, Thermal, Dielectric and Morphological Properties of Polyoxymethylene/Silica

Nanocomposites", *Journal of Nanoscience and Nanotechnology*, 18, (2018), 4963-4970.

10. Matteo, Tonezzer*; Dang, Thi Thanh Le; Salvatore, Iannotta; Nguyen, Van Hieu. "Selective discrimination of hazardous gases using one single metal oxide resistive sensor", *Sensors and Actuators B*, 277, (2018), 121-128.
11. Chu, Thi Quy; Nguyen, Xuan Thai; Nguyen, Duc Hoa*; Dang, Thi Thanh Le; Chu, Manh Hung; Nguyen, Van Duy; Nguyen, Van Hieu**. "C_2H_5OH and NO_2 sensing properties of ZnO nanostructures: correlation between crystal size, defect level and sensing performance", *RSC Advances*, 8, (2018), 5629-5639.
12. Kien, Nguyen; Nguyen, Duc Hoa*; Chu, Manh Hung; Dang, Thi Thanh Le; Nguyen, Van Duy; Nguyen, Van Hieu. "A comparative study on the electrochemical properties of nanoporous nickel oxide nanowires and nanosheets prepared by hydrothermal method", *RSC Advances*, 8, (2018), 19449-19455.
13. Pham, Long Quang; Nguyen, Duc Cuong; Tran, Thai Hoa; Hoang, Thai Long; Chu, Manh Hung; Dang, Thi Thanh Le; Nguyen, Van Hieu. "Simple post-synthesis of mesoporous p-type Co3O4 nanochains for enhanced H_2S gas sensing performance", *Sensors and Actuators B*, Volume 270, (2018), 158–166.
14. Vo Thanh, Duoc; Dang, Thi Thanh Le*; Nguyen, Duc Hoa**; Nguyen, Van Duy; Chu, Manh Hung; Hugo, Nguyen; Nguyen, Van Hieu. "New design of ZnO nanorods and nanowires based NO2 room temperature sensors prepared by hydrothermal method", *Journal of Nanomaterials*, 2019, Article ID 6821937, 9 pages.
15. Tran, Thi Ngoc Hoa; Nguyen, Duc Hoa*; Nguyen, Van Duy; Chu, Manh Hung; Nguyen, Van Toan; Dang, Thi Thanh Le; Nguyen, Huy Phuong; Nguyen, Van Hieu. "Effective H_2S sensor based on SnO_2 nanowires decorated with NiO nanoparticles by electron beam evaporation", *RSC Advances*, 2019, accepted.

In: ZnO Nanostructures
Editor: Dana Crawford
ISBN: 978-1-53616-773-3

Chapter 4

Luminescence and Catalytic Properties of ZnO and Its Heterostructure: Effect of Morphology and Defects

Puja Bhattacharyya and Chandan Kumar Ghosh*
School of Materials Science and Nanotechnology, Jadavpur University, Jadavpur, Kolkata, India

Abstract

ZnO, a well-abundant material, has a wide range of potentiality in various optoelectronic devices such as light emitting diode, solar cell, transducer, waveguide, sensor etc., attributed to its environmental biocompatibility, chemical stability, suitable band gap (~ 3.37 eV), electron mobility etc. Other important properties which make it attractive in diverse field are piezoelectricity, excitonic (~60 meV), photostability, electron donating and accepting property, photocatalytic etc. In general, all optoelectronic applications of ZnO primarily depend on band to band near

* Corresponding Author's Email: chandu_ju@yahoo.co.in.

edge and defect mediated electronic transitions. Defects, classified as intrinsic defects and extrinsic defects, can be varied easily in ZnO by synthesis conditions; hence by controlling defects, ZnO can be tailor made for various optoelectronic applications. Intrinsic defects get formed from either interstitial or deficiency of Zn and or O, while extrinsic defects are related with doping with other elements *viz.* P, Mn, Ni, Al, S, N etc. On the basis of the position of defect levels with respect to conduction and valence band edges, they are classified into deep level defects and shallow defects. In general, shallow defects contribute to electrical conductivity, while deep level defects significantly effects on optical transition giving luminating property and photocatalytic activity of ZnO. In recent time, it has been emphasized that optoelectronic properties of ZnO is modified widely in its nanostructured form and much efforts have been given to understand the optoelectronic behaviour of ZnO, followed by device fabrication. Various techniques like hydrothermal, solvothermal, sol-gel, chemical bath deposition, physical vapour deposition, atomic layer deposition, pulsed laser deposition have been adopted to generate different nanostructure of ZnO such as nanorods, nanoplate, nano-flower, hedge – hog etc. In general, all these synthesis processes introduce different kinds of defect, morphology etc. that impact differently on optoelectronic properties of ZnO nanomaterials, particularly in fabricating light emitting diode. Thus, to gain insight into defect mediated application aspects of ZnO nanostructured materials, their correlation with the defect structure is highly essential and it is being elaborated within this chapter. ZnO nanostructures are also being used in degradation of toxic dyes by photocatalytic process, ascribed to suitable energy position of conduction and valence band edges to generate reactive oxygen species. Here also defects play an important role to inhibit electron – hole recombination, which successively increases catalytic efficiency. In this context, it may be stated that stability of the defects as determined by thermodynamics is important. Nowadays, efforts are given to tune optoelectronic properties of ZnO by forming its heterostructure, where photo-excited electron transfer across the junction plays crucial role in modifying the optoelectronic properties of ZnO in light emitting devices and catalytic applications. This present chapter also deals with synthesis protocols of ZnO heterostructures along with its role in these optoelectronic applications. In this context, it may be stated primarily that charge separation across the heterostructure which causes less recombination of photo-generated electron and hole tunes plays important role to tune emission and catalytic efficiency. In brief, present chapter highlights thermodynamic stability, and correlation among morphology, defects, heterostructure with luminating and catalytic properties.

1. Introduction

ZnO nanostructures (~ 5 – 100 nm) have triggered attention of researchers due to their potentiality in various optoelectronic devices, attributed to their fascinating and unique properties. It possesses three types of unit cell structure: rock salt, zinc blende and hexagonal wurtzite. In rock salt ZnO, each Zn or O atoms are surrounded by six alternative atoms in its unit cell. It exists only at high temperature and is found to be unstable in epitaxial form. Zinc blende, on the other hand, has cubic unit cell structure, but its synthesis remains a challenge till date. In comparison to rock salt or zinc blende, hexagonal, wurtzite ZnO comprising of four Zn ions at corners of the tetrahedral coordinates and O ions at the centre is most stable at normal atmospheric condition and has got most industrial attention. Hexagonal ZnO has band gap (E_g) ~ 3.3 eV, good transparency in visible region, moderate electrical conductivity, high thermal and mechanical stability etc., thus it suits for various optoelectronic devices such as scintillation detectors, UV emitting devices, UV light detectors, plasmonic laser, photovoltaic cells, sensors etc. ZnO is very much cost-effective and biocompatible too. Due to intense emission in visible region, ZnO fits itself as a potential material for various optical devices including flat panel display, light emitting diode (LED) etc. In recent time, LED technology is being widely used in automotive headlamp, advertisement festoon, traffic signal, aviation lighting etc. as this technology exhibits several advantages over conventional fluorescent and incandescent light such as free from heavy metal (Hg), high conversion efficiency of electrical energy into light energy, long life time, low power consumption etc. ZnO possesses exciton binding energy (~ 60 meV) higher than room temperature thermal energy (~ 25 meV), thus it has the potentiality in exciton-type short-wavelength optoelectronic devices, particularly in ultraviolet laser. Therefore, it may be summarized that ZnO nanostructures cover wide range of applications in various optoelectronic fields and the opportunity is growing rapidly. It has been noticed that the emissions from ZnO nanostructures significantly depend on their defects and morphology. In this context, it may be stated that various morphologies such as nanowires, nanotetrapods, nanoribbons, nanorods, nanohelix, nanodisks etc. are easily being synthesized by several low-cost, simplest techniques like hydrothermal, micro-emulsion, spray pyrolysis, ball milling etc. Here, uncoordinated surface atoms play an important role in the optical properties of ZnO nanostructure materials and in this chapter, optical property, particularly luminescence property, along with the mechanism will be briefly discussed.

Pollution becomes major threat of present decade as earth's ecological balance is going down by ever increasing pollutants like microbes, toxic dyes, carcinogenic metals etc. In general, discharge of untreated industrial hazards into water, land or air causes acute and chronic detrimental effect to the ecological system. For example, high level of toxic Cr^{+6} in water, disposed into from tannery industries, impacts adversely on marine life and human life too. Excess usage of pesticide increases toxicity of water and land and enters into the food chain. Nowadays, pollutants are worldwide challenge as their toxic effect is crossing borders of almost each and every nation. Therefore, urgency of removing pollutant is in high alert for our healthy living condition. Among various techniques to reduce environmental toxicity, catalytic degradation has got attention due to its simplest mechanism and easy handling process. Among different catalytic processes such as photocatalytic, electrocatalytic, thermocatalytic etc., photocatalytic has gained attention due to its cost-effectiveness. Here, light energy is utilized to initiate chemical reaction which degrades toxic pollutant into less toxic ingredients. Among various photocatalysts, ZnO nanostructures have been identified as an efficient photocatalyst and the photocatalytic activity significantly depends on its shape, size, defects etc. and causes degradation of environmental pollutants including toxic dyes, carcinogenic metal etc. In addition, it has also been noticed that microbes are becoming resistive to conventional anti-biotic, thus alternative protocols to fight against harmful microbes becomes another challenge where ZnO plays a prominent role due its anti-bacterial activity, attributed to its catalytic property. In this chapter, we'll emphasize mechanism of photocatalytic activity of ZnO nanostructures along with its correlation with defects, morphology etc.

2. LUMINESCENCE PROPERTY OF ZNO

Luminescence from ZnO, the emission of light under external stimuli, encompasses a striking potentiality in recent past as the emission covers a wide range of colours – near UV, blue, violet, green, yellow, orange and red etc. with significant rendition. Stimulations giving the emission may be another light radiation, cathode ray, temperature, external electric field etc. and accordingly they are classified as photoluminescence, cathodoluminescence, thermoluminescence and electroluminescence (Das et al. 2017). It has been identified that a delay time (10^{-12} seconds - few seconds) exists between excitation and emission processes. Based on the delay time, luminescence (schematically shown in Figure 1(a)) is divided into (i) fluorescence where emission ceases immediately after switching off the stimulation and (ii) phosphorescence where afterglow persists for longer time. Quantum mechanically, electronic states are classified as singlet and triplet state according to their resultant angular momentum (S). Singlet state, occupied by a pair of electrons, doesn't possess angular momentum (S = 0), while triplet state carries net momentum (S = 1). Transition between two singlet states gives fluorescence, whereas phosphorescence originates from electronic transition between triplet and singlet states.

Most importantly, ZnO is found to be one of the rarest materials exhibiting wide variation in its luminescence including fluorescence, phosphorescence, cathodoluminescence and electroluminescence depending on size, morphology, defects. Due to such versatile characteristics, ZnO appears as a potential material for many optoelectronic devices. In this regard, it may be stated that most of the semiconductors don't show luminescence property, i.e., materials having luminescence property exhibit few salient features. Briefly, when an electron gets excited, de-excitation occurs either by emission of energy in the form of light, called radiative transition, or by release of energy through thermal vibration, called non-radiative transition. The picture becomes clearer from configurational coordinate versus energy diagram (shown in Figure 1(b)) where quantized energy states of electron and lattice for both ground and excited states are plotted. For materials having weak coupling between electron and lattice, absorption and emission occur vertically and they exhibit luminescence.

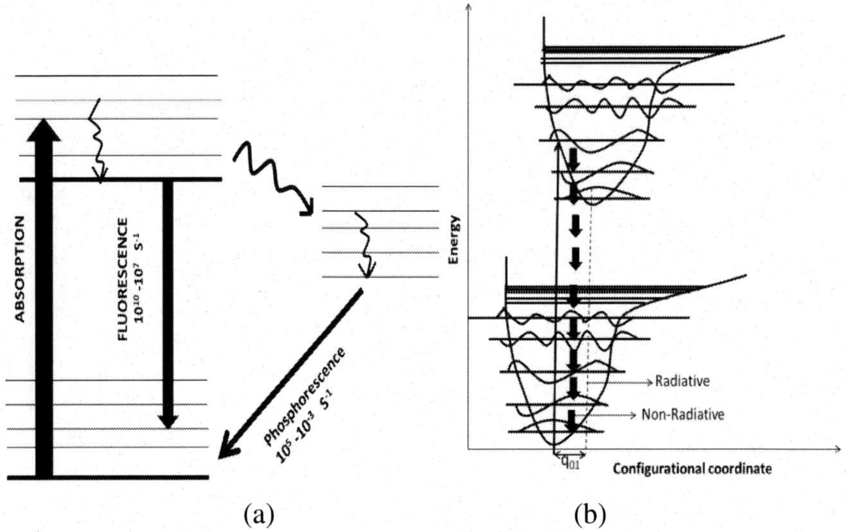

Figure 1(a). Fluorescence and Phosphorescence (b) Schematic representation of radiative and non-radiative transitions in energy vs. configurational coordinate diagram.

In contrast, materials with strong electron – lattice interaction possess non-radiative transition as excited electrons easily get relaxed through lattice vibration and in this case non-radiative transition dominates over radiative transition. Most of the materials follow later process, thus they appear as non-luminating. Generally, luminescence centers don't coincide and emission occurs at higher wavelength in comparison to exciting wavelength, this deviation is known as Stokes' shift. In this context, it may be stated that Stokes' shift provides a measure of interaction between excited electron and lattice. Though most of the luminating materials are insulator, but ZnO is the rarest luminating material having semiconducting character, attributed to its delocalized conduction and valence band. Fluorescence properties of ZnO (Shown in Figure 2(a)) originate from either band-to-band radiative transition in UV region (360 – 390 nm) or defects induced transitions in visible region (400 – 800 nm). Sometimes, band-to-band transition, referred as near band edge (NBE) emission, is dominated by electron and hole pair quasi-particle, called exciton. In addition to fluorescence, ZnO also shows phosphorescence properties, attributed to trapping of electrons at defect sites. Here, it can be stated that ZnO nanostructures exhibit both fluorescence and phosphorescence properties depending on their morphology, size, defects etc. Origin of such tuneable emission along with mechanisms will be briefly discussed in the next section.

2.1. Photoluminescence Property of ZnO

2.1.1. Near Band Edge and Excitonic Emission

NBE emission generally falls in the UV region with lifetime of excited electrons in the sub-nanosecond range, suitable for high-speed optoelectronic devices like lasers, scintillators etc. In ZnO, NBE emission is often dominated by exciton which further improves candidature of ZnO in many optoelectronic applications. Main advantage of exciton is that it carries energy without carrying charge, but it only gets formed in well crystalline ZnO. Depending on type of charge carriers participating in the transition processes, excitons are classified as (i) free exciton (FX) where both electron and hole are free and (ii) bound exciton where one of the charge carriers is bound to defects or impurities.

Depending on symmetry of the valence band, involved in the excitonic transitions, FX appears to be anisotropic and possesses different intensity and characteristics in different directions (Ho et al. 2008). Polarized photoluminescence spectroscope is generally used to examine anisotropies in FX and is explained as follows: conduction and valence bands of ZnO mainly constitute from Zn 4s orbital with Γ_7^c symmetry and O 2p orbital that splits into three bands (A, B and C), referred as heavy hole band with Γ_9 symmetry, light hole band with Γ_7 symmetry and splitted band with Γ_7 symmetry (Shown in Figure 2(b), Birman J. L. 1959, Mang A, Reimann K,. Rübenacke St. 1995). In this context, it may be stated that valence band generally splits due to crystal-field and spin-orbit interaction. According to group theory, ground state of exciton (illustrated in Figure 2(b)) can be written in the following way:

$$\Gamma_7 \times \Gamma_9 \rightarrow \Gamma_5 + \Gamma_6 \quad (1)$$

$$\Gamma_7 \times \Gamma_7 \rightarrow \Gamma_5 + \Gamma_1 + \Gamma_2 \quad (2)$$

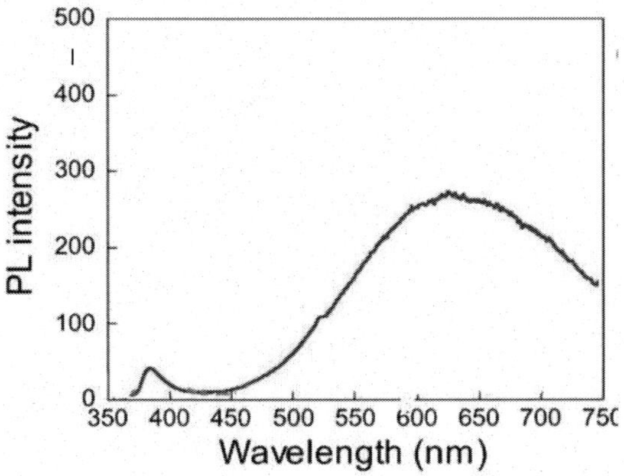

(a)

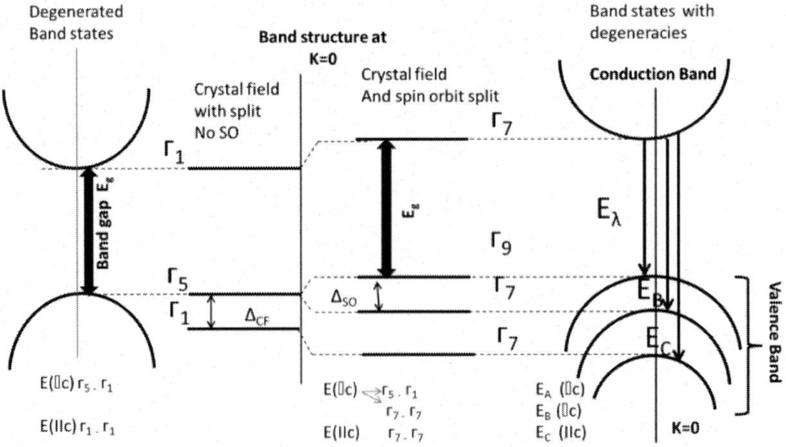

(b)

Figure 2. (a) Typical room temperature PL spectra of ZnO (Hsu et al. 2006) (b) Energy – momentum (E-k) diagram of ZnO.

where, Γ_1 andΓ_2 correspond to non-degenerate exciton ground states and Γ_5 and Γ_6 represent doubly degenerate states. According to selection rule of excitonic transitions, $\Gamma_5$and $\Gamma_1$are allowed for **E** ⊥**c** and **E** ∥ **c** respectively, whereas $\Gamma_6$and Γ_2 transitions are forbidden. Therefore, excitonic transitions are observed from all three splitted bands by σ-polarized light (**E**⊥**c,** and **k** ⊥**c**), but C-band exciton is found to be very weak in this case. For π-polarized light (**E** ∥ **c,** and **k** ⊥ **c**), C-band exciton is very strong, B-band exciton is moderate, while A-band exciton is forbidden. All these transitions have moderate intensity under α-polarized light (**E** ⊥ **c,** and **k** ∥ **c**). In this context, it may be stated that the polarization dependence of exciton is very prominent in ZnO nanowire (Jacopin et al. 2011), but it decreases with increase in temperature.

Defect or impurity of ZnO often acts as electron or hole trapping centre, thus they leads formation of bound excitons (Banerjee et al. 2017). Bound excitons, classified as neutral donor (D^0X)/acceptor (A^0X) exciton, charged donor (D^+X, $D^{++}X$)/acceptor (A^+X) exciton etc., predominantly forms within ZnO nanostructures. In this context, it may be stated that bound excitons are classified according to their charge, position of the defect or impurity states. Importantly, exciton can be easily tuned by its synthesis techniques, however it dominates at low temperatures whereas defects giving visible emission contribute at room temperature and in most of the cases, these two emissions are found to be complimentary (Bekeny et al. 2006). Thus to avoid overlapping from different emissions (discussed later) and to exclude temperature effects, excitonic experiments are performed at low temperature (4 – 100 K) region. Before elaborating excitonic transitions in various ZnO nanostructures, we give a typical overview on excitonic emission which consists of three peaks at 3.375, 3.360 and 3.333 eV, ascribed to FX, neutral donor bound exciton (D^0X) and acceptor bound exciton (A^0X) respectively. Intensity of the excitonic transition increases with decreasing particle size, whereas full width at half maxima (FWHM) shows opposite trend (Rodnyi et al. 2011).

Emissions between 3.34 – 3.31 eV, referred as two-electron satellites (TES, shown in Figure 3(a)), are observed in few ZnO nanostructures (Reynolds et al. 1998). The origin of TES is explained on the basis of hydrogen-like effective mass approach (EMA) in which energies of ground (1S) and excited states (2S and 2P) of an exciton (illustrated in Figure 3(b)) are obtained by solving Schrödinger's equation and TES is ascribed to the excitonic transitions of excited states. It has been experimentally observed that the energy differences between two consecutive states are exactly equal to ¾ of the energy of D^0X.

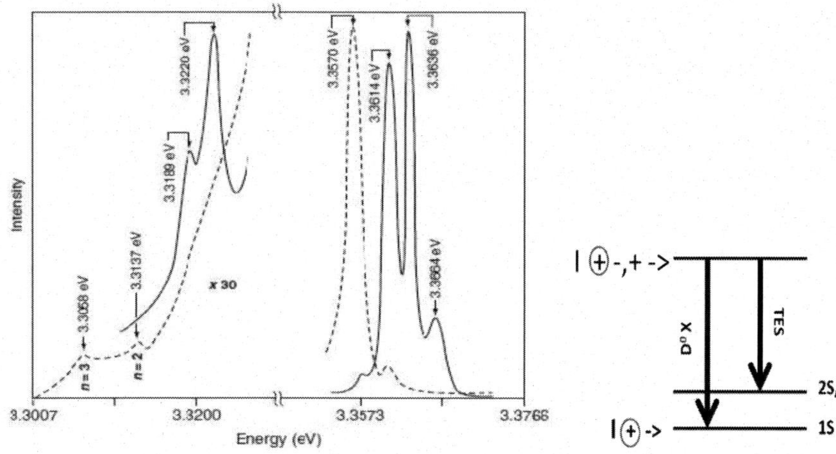

(a) (b)

Figure 3. (a) Two-electron transitions at 3.322 and 3.3298 eV, associated with D^0X. (Reynolds et al. 1998) (b) Schematic representation of the origin of satellite peaks in excitonic emission.

In general, bound excitons are characterized by their localization energy (E_L) which is defined as the energy difference between recombination of free exciton (E_{FX}) and bound exciton ($E_{(D/A)X}$) i.e., $E_L = E_{FX} - E_{(D/A)X}$. In this context, Haynes observed that E_L is proportional to the binding energies of the donors/acceptors those act as exciton trapping centre. This is known as Haynes rule (Haynes 1960). The trend $E_L < E_L(D^0X) < E_L(A^0X)$ is noticed for 'E_L' and is explained on the basis of effective masses of the charge carriers, participated in excitonic processes. In this respect, it can be stated that the charge carrier with higher effective mass exhibits higher localization energy. D^0X shows higher intensity than FX in various ZnO nanostructures due to presence of higher defect sites. This effect is found to be very much significant specifically in ZnO nanowire as intensity ratio ($\frac{I_{FX}}{I_{D^0X}}$) is measured to be ~ 0.03 (Zimmler et al. 2009).

In addition to TES, few excitonic emissions are also observed at higher (> 3.375 eV) as well as lower energy sides (< 3.375 eV) of FX due to interaction of exciton with other quasi-particles like phonon, polariton etc. As an example, ZnO nanorods, grown on Si substrate, exhibit a dominant D^0X emission at 3.362 eV at 4 K along with two additional emission peaks at 3.289 and 3.216 eV (Shown in Figure 4(a)). Additional emissions, referred as phonon replica, are assigned to the interaction of electrons with linear optical (LO) phonon (Bekeny et al. 2006). In general, electron – LO phonon interaction in ZnO, represented by Frölich interaction model, plays a significant role in FWHM of the emission spectrum too. In fact, FWHM is found to be proportional to this interaction.

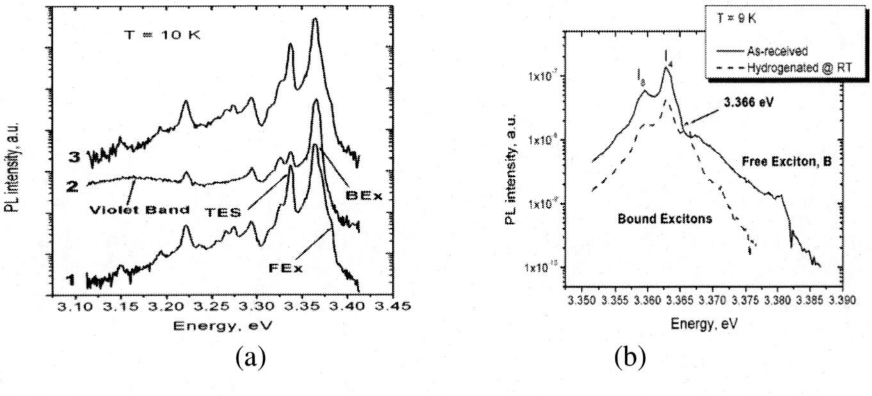

Figure 4 (a): Phonon-replica of bound excitonic emission, (b) Low-temperature PL in near-band edge region (Strzhemechny et al. 2003).

Excitonic peaks, appeared at higher energy side, originate from the interaction of exciton with polariton (Shown in Figure 4(b)). As polariton has anisotropy, so its influence on excitonic emissions also noticed to be anisotropic. Briefly, polariton sets a short range chemical potential which induces anisotropy and consequently fine-structure of excitonic emissions gets generated. Theoretically, polariton effects are studied on the basis of second order perturbation theory. According to this theory, energies (E_{1S},

E_{2S}, $E_{2P_{x,y}}$ and E_{2P_z} for 1S, 2S, $2P_{x,y}$ and $2P_z$ respectively) of the excitonic states are calculated and they are given by the equations 3(a –d):

$$E_{1S} = -R \quad (1 + 0.72\gamma^2) - \Delta E_{1S}^{ch} \qquad 3(a)$$

$$E_{2S} = -(R| \quad | \quad /4)(1 + 1.125\gamma^2) - \Delta E_{2S}^{ch} \qquad 3(b)$$

$$E_{2P_z} = -(R| \quad | \quad /4)(1 + 0.8\gamma^2) \qquad 3(c)$$

$$E_{2P_{x,y}} = -(R| \quad | \quad /4)(1 - 0.4\gamma + 0.5\gamma^2) \qquad 3(d)$$

where, ΔE_{1S}^{ch}, ΔE_{2S}^{ch} and γ represent chemical shift of 1S state, 2S state and correction constant respectively. 'R , defined as effective Rydberg constant, gets evaluated from equation (4):

$$R \quad = Ry\frac{m_e}{m_0}\frac{1}{\varepsilon_0^2} \qquad (4)$$

In this equation, Ry= 13.59 eV represents Rydberg constant, m_0 and ε_0 represent effective free electron mass and effective dielectric constant which is the geometric mean ($\varepsilon_0 = \sqrt{\varepsilon_0^{\|}\varepsilon_0^{\perp}}$) of parallel ($\varepsilon_0^{\|}$) and perpendicular ($\varepsilon_0^{\perp}$) component of dielectric constant. Here, m_e is given by equation (5):

$$\frac{3}{m_e} = \frac{2}{m_e^{\perp}} + \frac{\eta}{m_e^{\|}} \qquad (5)$$

In this expression, $m_e^{\perp}$ and $m_e^{\|}$ represent perpendicular and parallel electron – polaron effective mass respectively and constant (η) is calculated from the relation$\eta = \frac{\varepsilon_0^{\perp}}{\varepsilon_0^{\|}}$. Anisotropy constant ($\gamma$) is obtained from equation (6):

$$\frac{3\gamma}{m_e} = \frac{1}{m_e^{\perp}} - \frac{\eta}{m_e^{\parallel}} \tag{6}$$

Therefore, it may be stated that the anisotropy is mainly caused by anisotropic effective mass of the charge carrier, dielectric constant and 'γ'. Various excitonic emissions, caused by different intrinsic quasi-particle interaction, is schematically represented in Figure 5. In addition to intrinsic mechanisms, FX and NBE emissions are also tailored by several extrinsic techniques such as doping etc. As an example, doping with Bi, Cu, etc. causes red-shifted (Xu et al. 2007, Raji et al. 2017), while Mg doping gives blue shift (Singh et al. 2012) in the excitonic emissions. In addition, doping with heavy metal (Ga, In etc.), hydrogen inter-diffusion, annealing etc. are other techniques to tune excitonic emissions, particularly to enhance intensity of excitonic emission (Khomyak et al. 2013, Dev et al. 2011).

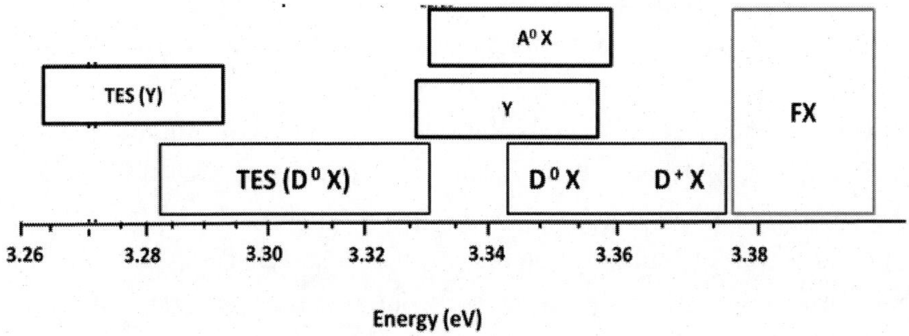

Figure 5. Block diagram of excitonic range.

In recent time, tuning of excitonic transitions by tailoring morphology appears to be very attractive as various morphologies of ZnO with different excitonic emission are easily achieved by several low cost synthesis techniques such as hydrothermal, sol-gel, co-precipitation etc. In addition, synthesis parameters like precursor concentration, pH, temperature, time etc. are often used to modify excitonic emissions. As an example, ZnO nanostructure, prepared at pH ~ 8 and 9, exhibits luminescence peak at 386 and 393 nm respectively. Redshift is assigned to competitive evolution of free exciton into bound exciton (Flores et al. 2014). In this context, it may be stated that UV to visible intensity ratio ($\frac{I_{UV}}{I_{visible}}$) also depends on synthesis methods as studied by Barbagiovanni et al. (Barbagiovanni et al. 2016). This research group has synthesized ZnO nanorods using three different techniques: vapour–liquid–solid method (VLS), chemical bath deposition (CVD) and hydrothermal and have found highest $\frac{I_{UV}}{I_{visible}}$ in VLS, followed by CBD and hydrothermal.

Another technique that has been recently developed to tune excitonic transitions and NBE emissions is hetero-structuring where metallic nanoparticles like Al, Au, Ag etc. get adsorbed on ZnO nanostructures (Wu et al. *2011*). In these heterostructures, excitonic transition gets changed predominantly by exciton – plasmon interaction. For example, low temperature NBE emission in ZnO nanorods which is dominated by surface defect bound exciton gets significantly enhanced by Ag coating. The enhancement is explained on the basis electron migration from Ag nanoparticle to ZnO nanostructure due to difference in their Fermi energies. Briefly, migrated electrons neutralize surface defects of ZnO and consequently enhance NBE emission (Cheng et al. 2010). In addition, Ag decoration on ZnO nanostructure supresses phonon replica and reduces lifetime of excited electrons, i.e., faster operation of optoelectronic devices can be achieved by ZnO/Ag heterostructure (Bhattacharyya et al.2018) in comparison to bare ZnO. But, completely opposite phenomenon is observed for ZnO/Pd heterostructure (Bera et al. 2011) where NBE decreases in the presence of Pd. After extensive research work in this field, now it has been realized that the size and distance between metallic nanoparticle and ZnO surface play a crucial role in tailoring NBE emission. For example, Chen et al. has observed an enhanced NBE emission due to 40 nm Au nanoparticle, whereas 15 and 75 nm Au nanoparticle quenches the emission (Chen et al. 2008). The effect of distance has been successfully corroborated by Zhao et al. when he observed significant enhancement (~ 120 times) of NBE emission in the presence of a dielectric spacer layer (HfO_2) between ZnO nanorods and Al nanoparticles (Zhao et al. 2018). It has been analysed here that dielectric layer causes localization of plasmon. Then this localized plasmon, called localized surface plasmon (LSPR), gets resonantly coupled with exciton and transfers energy by a new mechanism, called localized surface plasmon resonance energy transfer (LSPRET). In this context, it may be stated that non-radiative Förster energy transfer (FRET) mechanism predominates in direct contact heterostructures. Here, thickness of HfO_2 plays the crucial role in determining the dominance of these two competitive LSFRET and FRET processes. They have identified that the efficiency (η) of the non-radiative energy transfer can be expressed by the following equation (7):

$$\eta \frac{R_0^6}{R_0^6+d^6} \tag{7}$$

where 'R_0' and 'd' represent FRET distance and separation between ZnO and metal nanoparticle respectively. Expression indicates that η decreases with increasing thickness, i.e., LSFRET mechanism dominates at higher 'd'. But, the enhancement isn't observed beyond a critical thickness of the spacer layer. It has been experimentally verified that FRET dominates at HfO_2 thickness < 5 nm, whereas LSFRET plays role at thickness > 7 nm. Similar result is also obtained in ZnO/SiO_2/Ag heterostructure samples (Xiao et al. 2010). Clearer picture of this phenomenon gets emerged from finite difference time-domain (FDTD) simulation where local electric field around nanostructure in the presence of metallic nanoparticles is computed using discretized Maxwell's equation. Practically, this technique was employed in ZnO/SiO_2/Ag heterostructure to examine confinement of plasmonic field by SiO_2 spacer layer and the study reveals that the confined plasmonic field actually tailors NBE emission from ZnO nanostructure. In this context, it may be stated that such confinement of plasmonic field also reduces plasmonic losses; hence these heterostructures have strong potentiality in developing plasmonic laser (Oulton et al. 2009).

2.1.2. Defect Induced Visible Emission

In addition to UV, emission from ZnO in visible range is being widely utilized to develop various optoelectronic devices like white light emitting diode etc. In comparison to UV emission, more research-work has been carried out to understand the mechanism as lots of controversy existed about origin of the luminescence. Broad luminescence spectrum of ZnO in visible range consists of blue emission at 450 nm, green emissions at 500 and 550 nm, yellow emission at 585 nm and orange red emission at 600 – 750 nm and they are attributed to various defects such as oxygen vacancy (V_O), zinc vacancy (V_{Zn}), oxygen interstitial (O_i), zinc interstitial (Zn_i), antisite oxygen (O_{Zn}), surface defects etc. (Wei et al. 2007, John et al. 2011).

Blue emission from bare ZnO is assigned to electronic transitions between V_O/extended Zn_i and valence band or conduction band and O_i (Brahma et al. 2015). There are few reports illustrating that the blue shift of the green emission due to quantum confinement effect gives blue emission while violet – blue emissions are linked with V_{Zn}.

Origin of the green emission from bare ZnO nanostructure is controversial as many defect states likeV_O, V_{Zn}, O_i, Zn_i, O_{Zn}etc. are predicted to be involved here. After thorough investigations by several researchers, it is now accepted that green emissions at 500 and 550 nm originate from V_O and doubly charged oxygen vacancy (V_O^{++}, Rauwel et al. 2011). In this context, it may be stated that surface defects play an important role in creation of V_O^{++} defects according to the following mechanism: hole in the valence band of ZnO gets trapped at surface defects, followed by tunnelling into V_O and consequently V_O^{++} gets formed. Recombination of electrons in either shallow trapped states or conduction band with deeply trapped hole in V_O^{++} gives green emission at 550 nm (Dijken et al. 2000). The relation between V_O and green emission can be clearly realized from the following discussion (Jian et al. 2015): formation of V_O in ZnO can be described by the following defect reactions

$$\text{ZnO} \rightarrow V_{Zn}O + \text{Zn (g)} \tag{8}$$

$$\text{Zn (g)} \rightarrow ZnV_O \tag{9}$$

$$\text{ZnO} \rightarrow \frac{1}{2}O_2(g) + ZnV_O \tag{10}$$

And according to law of mass action, concentration ofV_O^{++} ($[V_O^{++}]$) can be expressed by the following equation:

$$[V_O^{++}] = [V_O]\,\frac{g_{V_O^{+}}}{g_{V_O}}\,\exp[\frac{E_d - E_F}{k_B T}] \qquad (11)$$

where, $g_{V_O}$and$g_{V_O^{+}}$ represent the degeneracy of respective defect sites. E_d, $E_F$are ionization energy of V_O^{++} and Fermi energy respectively. Equation (10) indicates that $[V_O^{++}]$ can be increased either by increasing $[V_O]$ or by decreasingE_F. First approach significantly depends on the annealing condition, i.e., green luminescence can be easily tailored by annealing in reducing or oxidising atmosphere, while later is sensitive to concentration of charge carrier. Specifically, annealing in N_2, H_2 atmosphere increases intensity of green emission, while annealing in O_2 atmosphere reduces it. Annealing mostly creates high-level surface and sub-surfaceV_O those also act as hole trapping centre, thusV_O^{++} gets formed. V_O^{++} is found to be more sensitive to annealing in comparison to V_O and can be explained in the following way (Das et al. 2017): according to Kroger-Vink notations, formation of V_O can be expressed by equation (12)

$$O_O = \frac{1}{2}O_2 + V_O \qquad (12)$$

where $O_O$presents oxygen at an oxygen sites and concentration of V_O ($[V_O]$) can be written as $[V| \quad |O] \propto (p_{O_2})^{\frac{-1}{2}}$. Similarly, formation of V_O^{++} can be described by equation (13):

$$O_O = \frac{1}{2}O_2 + V_O^{+++2e'} \qquad (13)$$

Charge conservation principle illustrates, so can be written as. Controlling of green emission by tuneable E_F is achieved from ZnO heterostructures. It has been realized that heterostructures with Au, Ag, Al reduce intensity, while heterostructure with carbon based material such as graphene enhances the intensity. Such phenomenon is attributed to the electron migration depending on work function of the respective materials. Briefly, heterostructure of ZnO with materials of higher work function gives enhanced intensity. Simple calculation shows that the enhancement of in heterostructures with respect to of bare ZnO nanostructures is given by equation (14):

$$= e^{x} \tag{14}$$

where 'x' (in eV) represents the change in E_f. Origin of the green emission becomes clearer from PL excitation spectrum (PLE) and time resolved photoluminescence (τ-PL) investigations, carried out by Camarda et al. (Camarda et al. 2016). PLE spectrum which consists of a sharp threshold at ~ 3.5 eV and flat spectrum at higher energies suggests the nanostructures have to be excited by energy > 3.5 eV. Time decay curve, analyzed in two different time regions (20ns – 40μs and 40μs – 1ms) according to the relation$I \propto t^{-\alpha}$, indicates existence of two different decay processes with α = 1.05 and 1.8. In general, α = 1.00 – 1.50 is related with recombination of trapped electrons when they reach to randomly distributed recombination center by tunneling process, while α ~ 2 is ascribed to the bimolecular recombination mechanism of a balanced donor – acceptor pair (DAP). Thus, it may be stated that initial recombination process is driven by tunneling of trapped electrons, whereas after a long time, electrons and holes come out of the trapping centers and the recombination mechanism gets transformed into bimolecular type. Lifetime of electrons giving green emission is found to be ~ μs for most of the ZnO nanostructures, i.e., they exhibit phosphorescence property (Das et al. 2017). But the phosphorescence appears to be very sensitive on annealing, i.e., lifetime of the excited electron giving green luminescence significantly depends on annealing that qualitatively can be described as follows: generally, lifetime of an electron in excited state can be divided into radiative (τ_R) and non-radiative (τ_{NR}) lifetime. According to DAP formalism, radiative recombination rate $W_R\left(\frac{1}{\tau_R}\right)$ can be expressed by $W_R \propto \left|\int \psi_f P \psi_i dv\right|^2$, where ψ_i (ψ_f) and 'P' represent wave function of bound electron (hole) and momentum operator respectively. $'W_R'$gets simplified into the following form for a loosely bound electron – hole pair system:

$$W_R(r) = W_R e^{-2r/a_B} \tag{15}$$

where ‘r’ represents delocalization of electron – hole pair and ‘a_B’ is the Bohr radius. Intense green emission indicates very high and $[V_O$ and it is assumed that an impurity band gets formed from defect states. Aannealing in O_2 atmosphere reduces and $[V_O$, hence delocalization of the electron – hole pair gets decreased and successively decrease in τ_R is noticed. Opposite trend, observed for τ_{NR}, is interpreted on the basis of electron – LO phonon interaction as described by Huang – Rhys S-factor according to following relation:

$$S = \sum_q \frac{|V_q^2|}{(\hbar\omega_{LO})^2} |\rho_q|^2 \tag{16}$$

where, $'V_q'$, $'\rho_q'$ and $'\hbar'$ represent strength of the carrier – LO phonon interaction, Fourier transform of the charge density and reduced Planck's constant respectively. Under DAP recombination mechanism, Huang – Rhys S-factor gets simplified into the following relation:

$$S = \left(\frac{1}{\epsilon_\infty} - \frac{1}{\varepsilon_0}\right)\frac{e^2}{a_h}\frac{1}{\hbar\omega_{LO}}\left\{\frac{5}{16}(1+\sigma) + \frac{a_h}{R}[(\Delta_e + \Delta_h)(1-\sigma^2)^{-3} - 1]\right\} \tag{17}$$

Where, $\epsilon_\infty(\varepsilon_0)$, a_h, $'\sigma'$and ‘R’ are the high (static) frequency dielectric constant, orbital radius for hole, ratio of orbital radii of hole to electron (= a_h/a_e and the distance between them respectively and the parameters (Δ_e and Δ_h) are given by

$$\Delta_e = \left[1 - 3\sigma^2 + (1-\sigma^2)\frac{R}{a_e}\right] exp\left(\frac{-2R}{a_e}\right) \quad (18)$$

And

$$\Delta_h = \sigma^4\left[3 - \sigma^2 + (1-\sigma^2)\frac{R}{a_h}\right] exp\left(\frac{-2R}{a_h}\right) \quad (19)$$

Annealing in O_2 atmosphere increases average distance between electron and hole from 'R' to 'R + ΔR'. Assuming ΔR<<R, first order changes (Δ'_e and Δ'_h) in Δ_e and Δ_h can be approximated to the following relations:

$$\Delta'_e = \Delta_e - \left[2(1-\sigma^2)\frac{R}{a_e} + 1 - 5\sigma^2\right]\frac{\Delta R}{a_e} exp\left(\frac{-2R}{a_e}\right) \quad (20)$$

and

$$\Delta'_h = \Delta_h - \sigma^4\left[2(1-\sigma^2)\frac{R}{a_h} + 5 - \sigma^2\right]\frac{\Delta R}{a_h} exp\left(\frac{-2R}{a_h}\right) \quad (21)$$

Equations (20) and (21) state that Δ'_e and Δ'_h decrease with increasing 'R', thus consequentlyτ_{NR} gets enhanced. In this context, it may be stated that V_0 forms singlet ground state (S = 0), but in excited state it transformed into triplet state (S = 1). Therefore, relaxation of an electron from conduction band into the defect states involves electronic transition between triplet to singlet states, thus V_0 acts as phosphorescence centre.

Yellow emission in ZnO nanostructures is assigned toO_i andZn_i, while O_{Zn} causes orange-red emission (Manzano et al. 2011; Ahn et al. 2009) while red emission is related with structural imperfections or lattice disorder due to Zn_i and excess oxygen (Djurišić et al. 2007). In this context, it may be stated that Zn_i and O_i may occupy either octahedral or tetrahedral environment, but O_i at tetrahedral site is found to be most stable. Though, ZnO shows strong potentiality as phosphor material for various optoelectronic devices, but controversy still exists about their stability which will be discussed in the next sections.

2.2. Cathodoluminescence Property of ZnO

Cathodoluminescence (CL), utilized widely in modern flat panel display technology, is the general optoelectronic phenomenon in which a phosphor material emits light due to electron impingement. This is also a powerful technique to investigate luminescence properties of any optically active materials. In comparison to PL investigation, this is more accurate and reliable technique to examine luminesce property and related mechanism since larger number of electron – hole pairs are being generated by inelastic collision with impinging electrons. It is general observation that incidenting electrons with energy 'E' generate ~ $\frac{E}{3E_g}$ number of electron – hole pair, i.e., luminescence intensity can be easily tuned by this process. Another advantage is that it can map distribution of luminescence centres laterally and vertically of any materials which specifically helps to understand defect distribution of optically active nanomaterials with non-uniform distribution in luminescence centres. Briefly, luminescence from surface defect gets activated by incidenting electrons of low energy, while energized electrons provide deep defect related luminescence. Similar to PL, it is well accepted that CL emissions in visible region originates from defect states and most importantly this emission gets easily controlled by the incidenting electrons, therefore we can say that various colours can be generated from a single material. Thus, CL appears not only as an efficient technique to examine defect states, but it also plays a significant role to develop technology of flat panel display. CL based display technology exhibits several advantage over earlier conventional technologies as old technologies were suffering from high cost, low resolution etc. In addition, materials with intense CL property are often used as phosphor material for field emission based flat panel display. Vacuum compatibility of the underneath phosphor material was the major challenge, but ZnO shows vacuum compatibility. However, for high resolution, reliable display panel, underneath phosphor materials must have high electrical conductivity, low surface recombination velocity, high current densities etc. Earlier, sulphur based materials were used for this purpose, but low electrical conductivity, stability etc., limit their usage. Importantly, ZnO satisfies all the characteristic features for field emission display and in addition, low excitation voltage reduces operational cost.

Before we elaborate CL in bare ZnO, we give a typical overview of CL from various ZnO systems. For example, low temperature (80 K) CL of ZnO thin film consists of three peaks at 3.37, 3.30 and 3.23 eV and a broad emission peak in the visible range (Phillips et al. 2004). First UV peak is assigned to bound exciton whereas others peaks are attributed to phonon replica of FX. In contrast to bulk ZnO, CL of nanocrystalline ZnO consists of five peaks, measured at 470, 550, 600, 680 and 770 nm, are assigned to various defects as schematically shown in Figure 7(b) (Park et al. 2015).

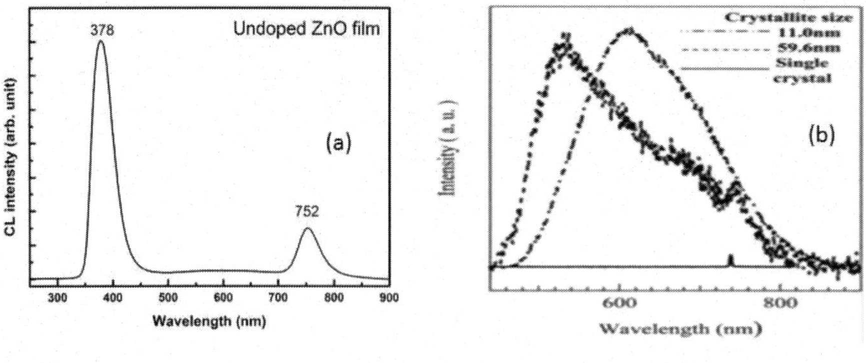

Figure 6: Typical cathodoluminescence spectra of (a) ZnO thin film (Ref: Highly transparent and conducting C:ZnO thin film for field emission displays, Zulkifli Z, Subramaniam M, Tsuchiya T, Rosmi M.S., Ghosh P., Kalita G., Tanemura M., RSC Adv., 2014, 4, 64763–64770) and (b) ZnO nanoparticle (Ref: Characterization of ZnO nanoparticles by resonant Raman scattering and cathodoluminescence, Yoshikawa M., Inoue K., Nakagawa T., Ishida H., Hasuike N., Harima H. Appl. Phys. Lett. 92, 113115 (2008)).

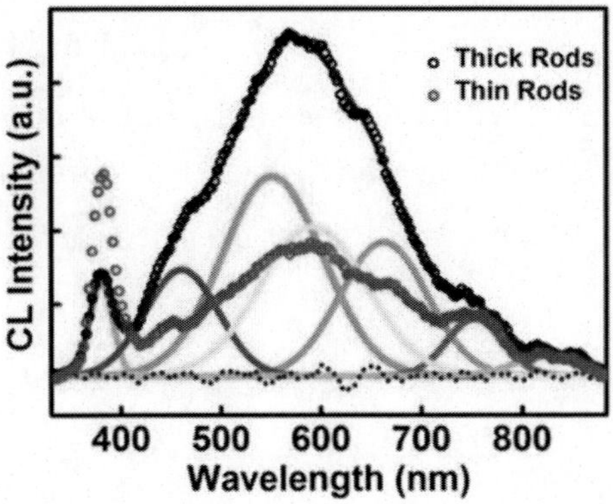

(a)

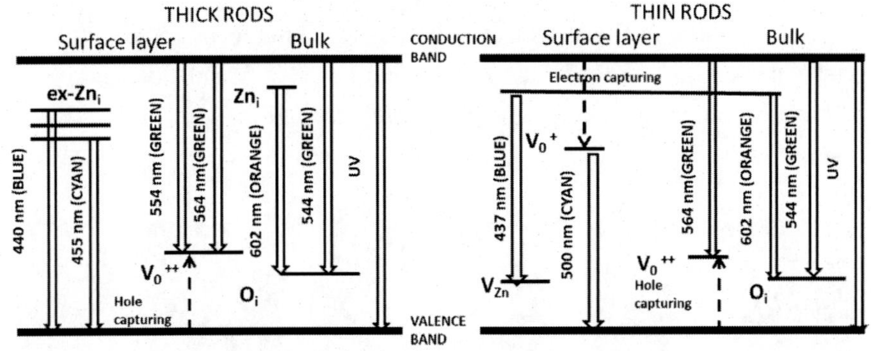

(b)

Figure 7. (a) Cathodoluminescence spectra of ZnO nanorods having different radius (Park et al. 2015), (b) Schematic diagram of the defect states.

In this context, it may be stated that CL property doesn't always follow monotonic feature of increasing visible intensity with decrease in size, rather it is found to be mostly dependent on defect concentration as the defects well below the surface also get excited by the impinging electrons. For example, CL spectra of ZnO nanorods of two different diameters, shown in Figure 7(a), illustrate that thick nanorods show higher intensity in visible range in comparison to thin nanorods. This trend is found to be opposite for emission in UV range. This is still controversial as different results are available. For example, Pan et al. has observed that CL emissions in UV as well as visible region decreases with decrease in the diameter of ZnO nanorods, prepared by vapour-phase transport and condensation method. Interestingly, intensity ratio of visible to UV emission ($\frac{I_{Visible}}{I_{UV}}$) is found to be increasing with decrease in the diameter (d) of the nanorods. But, it doesn't match in CL emission due to self-absorption effect which depends on 'd'. Based, on the self-absorption, the following relation has been deduced between $\frac{I_{Visible}}{I_{UV}}$ and 'd' (Pan et al. 2007):

$$\frac{I_{Visible}}{I_{UV}} = 0.643 \times \frac{0.2+1/d}{0.05+1/d} \times \left(4w_s \frac{1-\frac{t}{d}}{\left(1-\frac{2t}{d}\right)^2} \right) \times \frac{t}{d} w_b \tag{22}$$

where, 't' represents the effective volume beneath the nanorods surface entirely contributing to surface recombination. 'w_s' and 'w_b' are the weighted average of surface and core recombination respectively ($w_s + w_b = 1$).

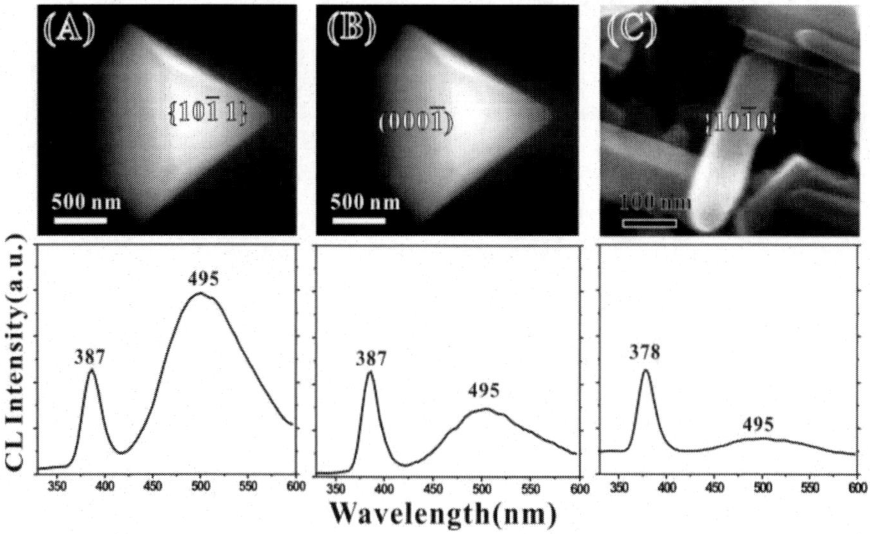

Figure 8. Facet dependent CL spectra (Zhou et al. 2007).

In this context, it may be stated that earlier controversy on the relation between oxygen defects and green emission has been successfully established by Zhou et al. using CL technique (Zhou et al. 2007). Briefly, they have identified highest green emission from $\{10\acute{1}1\}$ surface when they were examining CL spectrum from $\{10\acute{1}1\}$, $\{000\acute{1}\}$, $\{10\acute{1}0\}$ face exposed ZnO nanostructures and the interpretation was given as follows: in hexagonal ZnO, {0001} and $\{10\acute{1}1\}$ surfaces are polar, while $\{10\acute{1}0\}$ surface appears to be non-polar. Non-polar surface being electrically neutral has low surface energy, hence low defect concentration. Therefore, it provides low intensity for green emission. However, on {0001} polar surface, an O atom bonds with three underneath Zn atoms, whereas an O atom bonds with four Zn atoms in bulk ZnO, thus an O atom on {0001} surface carries 2/3×charge of an oxygen atom extra charge. Therefore, V_O gets formed during minimization of surface energy by reconstructing {0001} surface, hence it possesses moderate intensity for green luminescence. On the other hand, on $\{10\acute{1}1\}$ surface, an O atom has surplus charge ~ 5/6 × charge of an oxygen atom, thus $\{10\acute{1}1\}$ surface possesses highest polarity and consequently it has highest surface energy and V_O.

2.3. Thermoluminescence Property of ZnO

Thermoluminescence (TL) in which a phosphor material emits light in the presence of temperature is another method to study defects, particularly defects in nanostructure material. In this context, it may be stated that the defect-type, their energy, stability, defined by activation energy of the defect (E_a), is obtained from TL investigations where temperature dependence of the number of emitted photons due to recombination of excited charge carriers from trap centers with opposite charge carriers at recombination centres is measured and analysed. But, in comparison to PL or CL, comparatively less experiment has been carried out on TL property of ZnO. Here, we give brief overview about TL properties which can be divided into two regions (i) above room temperature and (ii) below room temperature. Above room temperature, two TL peaks at 420 and 490 K (typically shown in Figure 8) with E_a ~ 0.8 and 1.2 eV are observed from ZnO nanophosphor (Pal et al. 2006) while ZnO nanorods exhibit two TL peaks at 371 and 463

K with E_a ~ 0.54 and 0.89 eV. In the lower temperature region, TL spectrum of polycrystalline ZnO consists of five peaks at 112, 129, 144, 156 and 172 K (Connolly et al. 1968), while that of single crystalline ZnO corresponds to two peaks at 150 and 182 K.

Primary focus of TL spectroscopy is to measure E_a for different defect states and it is being obtained by fitting TL glow curve (S(T) curve) which plots total light emission as a function of temperature. In this context, it may be stated that non-linear least-square Marquardt fitting method is utilized to fit the curve. Kinetics of the process is given by the following relations (Chen et al. 1997):

$$S(T) = S_0 \left[\frac{-sk_BT^2}{\beta E} exp\left(\frac{-E}{k_BT}\right)\left(1 - \frac{2k_BT}{E}\right)\right] \quad (23)$$

$$S(T) = S_0 \left[1 + \frac{sk_BT^2}{\beta E} exp\left(\frac{-E}{k_BT}\right)\left(1 - \frac{2k_BT}{E}\right)\right]^{-1} \quad (24)$$

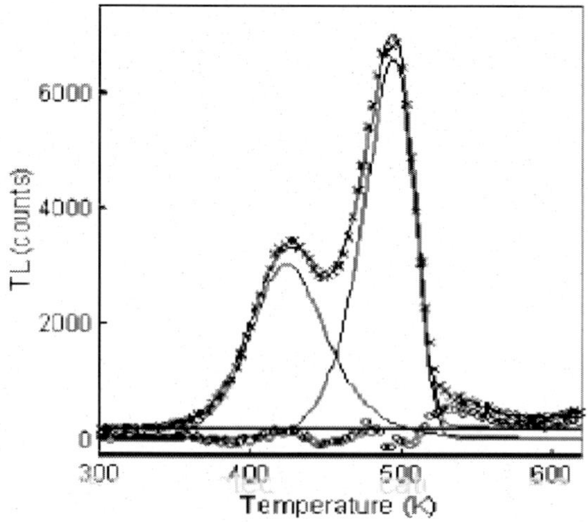

Figure 9. Typical TL signal from ZnO (Pal et al. 2006).

Where, s, β and S_0 represent frequency factor, heating rate, total TL signal, emitted during heating process respectively. Equation (23) and (24) stand for first order and second order processes. In this context, it may be stated that a lots of controversy existed about origin of green luminescence from ZnO, but TL spectroscopy resolves that green emissions mostly originate from V_O^{++}andV_O. In addition, TL spectroscopy illustrates that V_O^{+} being very much unstable easily decays into V_O. In contrast to bare ZnO, doped ZnO exhibits additional interesting features, related to defects. For example, bare ZnO possesses two TL emission peaks at 420 and 490 K with corresponding activation energy 0.8 and 1.2 eV, while 5% Yb doped ZnO displays a single peak at 480 K with activation energy ~ 1.19 eV (Pal et al. 2006). In this context, it may be mentioned that like other thermoluminescence materials, ZnO also have potential use as a radiation dosimeter to estimate ionizing efficiency of β-irradiation, γ-irradiation field.

2.4. Electroluminescence Property of ZnO

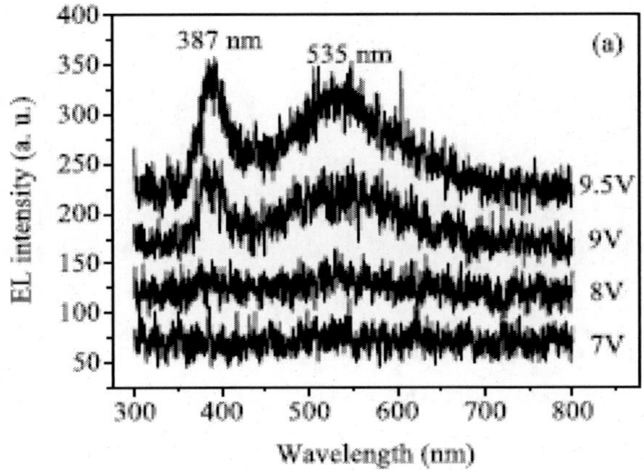

Figure 10. Electroluminescence spectra of n–ZnO/p–Si (Sun et al. 2006).

In electroluminescence, emission originates from a p – n junction due to passage of a current. ZnO has also been identified as an important material for electro-luminating applications, particularly in LED application, since various colours can easily be generated from ZnO just by varying defects. Additionally, it exhibits low loss and operational voltage due to its semiconducting character. Among different nanostructures, ZnO nanorods have been studied to exhibit best potentiality. Figure 10 illustrates a typical electroluminescence spectrum of ZnO nanorods. In this regard, p – n junction has been formed with n-type ZnO and various p-type materials such as p-GaN (Bano et al. 2017), p-Si (Raut and Rao 2008), while use of Au/ZnO heterostructure in contrast to bare ZnO enhances electroluminescence intensity (Yao et al. 2016).

3. Catalytic Property of ZnO

3.1. Thermodynamics and Kinetics of the Photocatalytic Process

After discovery of Honda – Fujishima effect in 1972, photocatalyses has been identified as one of the environmental friendly, cost-effective technique for H_2 generation from water, CO_2 reduction, toxic pollutants removal, bacterial degradation etc. Here, an electron gets excited into conduction band leaving a hole in valence band in the presence of a light with energy $>E_g$, then these photogenerated charge carriers take part in the photocatalytic process. Previously, high E_g materials were used, i.e., only UV portion of the sunlight was utilized for catalyses. In this regard, ZnO nanostructures have been found as an efficient photocatalyses due to high surface area, high absorption coefficient of light in UV range, less electron – hole recombination, high charge carrier mobility etc. As photocatalytic is a surface phenomenon, therefore it significantly depends on shape, size, exposed facet etc. of ZnO nanostructures. To use broader sunlight spectrum, presently efforts are given to develop visible light driven ZnO photocatalyst. It is the triumph that surface defects, doping etc. make ZnO nanostructures as visible light driven photocatalyses. Few other techniques such as heterostructure formation etc. also exist in this regard. Before detailed discussion about photocatalytic process of various ZnO nanostructures and heterostructures, it is highly essential to explain the fundamental principles of photocatalytic process.

It is the general phenomenon that photocatalytic reaction rate increases with temperature, thus light and heat simultaneously influence the photocatalytic process. However, Arrhenius-type temperature dependence of the reaction rate indicates that heat only initiates the reaction. Hence, proper understanding of the photocatalytic process is very much essential and can be represented in the following way: due to photo-excitation by photon having energy $>E_g$, an electron makes transition from valence band into conduction band leaving a hole in valence band. Electron in conduction band ($e^-_{CB,ZnO}$) of ZnO gets transferred to water dissolved $O_2$to form superoxide ion (O_2^-), similarly hole in valence band ($h^+_{VB,ZnO}$) reacts with H_2O and OH^- to form hydroxyl radical ($\bullet OH$) as represented by the following equations:

$$e_{CB,ZnO}^{-+O_2 \to O_2^-} \tag{25}$$

$$h_{VB,ZnO}^{++OH^{-} \wedge H_2O \to \bullet OH} \tag{26}$$

Generated O_2^- and $\bullet OH$ are highly reactive and cause degradation. In this context it may be stated that amongst large number materials, only few of them show photocatalytic activity as they exhibit special features. In this context, it may be stated that the oxidation of OH^- or $H_2O$and reduction of O_2 requires redox energy of -0.20 and +2.20 eV on normal hydrogen electrode (NHE) scale. Therefore, to generate O_2^- by $e^-_{CB,}$, conduction band must have energy lower than -0.20 eV, while for $\bullet OH$ generation by h^+_{VB}, valence band must have energy greater than +2.20 eV. Therefore, the materials having these characteristic energies for conduction and valence band exhibit photocatalytic reaction and most importantly ZnO satisfies this criterion. In Figure 11, we have represented band energies of few photocatalysts.

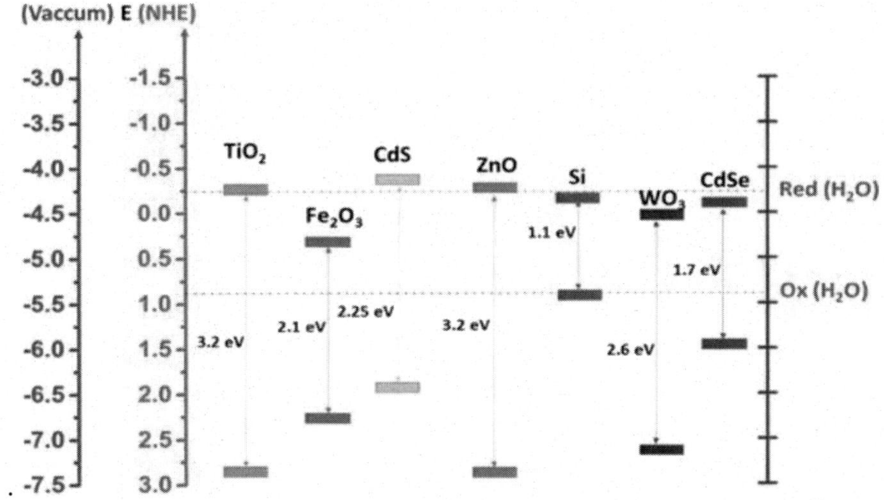

Figure 11. Conduction and valence band energies of few photocatalysts.

In this context, it may be stated that the photocatalytic process in any material is very complicated as it gets influenced by several parameters such as carrier generation, carrier transport, carrier recombination, carrier interfacial transfer etc. Importantly, ZnO exhibits strong potentiality, since it possesses high carrier mobility, high absorption coefficient and easy interfacial charge transfer path. Although, the above mentioned mechanism appears to be well-accepted, but it neither describes the driving force behind the process nor distinguish the contribution from heat and light. Proper mechanism can be understood in the following way: after interacting with lattice, $e^-_{CB,ZnO}$ and $h^+_{VB,ZnO}$ get relaxed and form quasi-equilibrium states. Quasi-equilibrium states at temperature 'T' can be represented by respective quasi-Fermi levels of electrons (F_e) and holes (F_h) according to the following relations (Liu et al. 2014):

$$F_e = E_C + k_B T ln \frac{n}{N_C} \tag{27}$$

$$F_h = E_V - k_B T ln \frac{p}{N_V} \tag{28}$$

where, E_C (N_C) and E_V (N_V) represent the energy (density of states) of conduction band minima and valence band maxima respectively, 'n' and 'p' are the concentration of electron and hole. Under light irradiation, thermodynamic driving force responsible for photocatalytic reaction can be expressed by the following equation:

$$\Delta G = G_{dark} - G_{light} = -|F_e - E_h| = -E_g - k_B T ln \frac{np}{N_C N_V} \quad (29)$$

In case of non-degenerate semiconductor, temperature induced electron (n_0) and hole (p_0) are related by the well-known equation (30):

$$n_0 p_0 = N_C N_V e^{-E_g/k_B T} \quad (30)$$

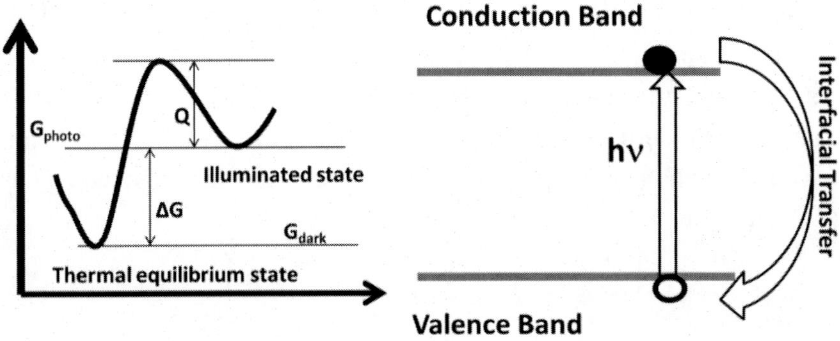

(a) (b)

Figure 12. (a) Diagram of thermal equilibrium state and illuminated state (b) Interdacial electron transfer.

Hence, $\Delta G = 0$ under thermal equilibrium condition, i.e., thermal energy can't lead photocatalytic process. However, under light irradiation, $F_e > E_h$, hence $\Delta G < 0$, i.e., photocatalytic process gets started, represented schematically in Figure 12 (a and b). Efficiency of the photocatalytic activity primarily depends on concentration of $e^-_{CB,ZnO}$ and $h^+_{VB,ZnO}$ and interfacial charge transfer rate (k_{IT}) which can be described by the following Arrhenius-type equation:

$$k_{IT} = \delta_0 exp^{\frac{-Q}{k_B T}} \tag{31}$$

Where, δ_0 is the pre-exponential constant and 'Q' represent the activation energy corresponding to interfacial electron transfer.

In this context, it may be stated that not only dynamics, kinetics also plays a vital role to determine efficiency of the photocatalytic process. In general, two types of kinetics exist here, one is first order and other is second order. Former kinetic process, referred as Langmuir – Hinshelwood model, deals with pollutant degradation where degradation rate $\frac{d[C(t)]}{dt}$ is proportional to its instantaneous concentration $[C(t)]$ as expressed by the following equation:

$$\frac{d[C(t)]}{dt} = -k[C(t)] \tag{32}$$

where, 'k' is the proportionality constant. According to equation (31), $[C(t)]$ follows the relation$[C(t)] = [C| \quad |0]exp(-kt)$, where $[C| \quad |0]$ represents initial concentration of pollutant. In other process, pollutant degradation rate $[C(t)]$ is described by the following relation

$$\frac{d[C(t)]}{dt} = -2k[C(t)]^2 \tag{33}$$

In this case, $[C(t)]$ follows the following relation$\frac{1}{[C(t)]} = \frac{1}{[C| \quad |0]+kt}$. Degradations by ZnO nanostructures mostly follow first order kinetic model.

3.2. Classification of the Photocatalytic Activity of ZnO

Photocatalytic activity of ZnO, divided into (i) indirect degradation and (ii) direct degradation, are summarized below:

(i) Indirect Degradation

Indirect degradation mechanism involves four steps: (a) photoexcitation (b) ionization of water, (c) oxygen ionosorption, (d) protonation of superoxide. In the first step, electron in conduction band and hole in valence band are generated by light irradiation with energy greater than E_g according equation (34):

$$ZnO + h\nu \rightarrow e_{CB,ZnO}^{-+h_{VB,ZnO}^{+}} \quad (34)$$

Ionization of water happens when $h_{VB,ZnO}^{+}$ reacts with water, adsorbed on the surface of nanoparticles and the ionization leads $\bullet OH$ generation as expressed by equation (35) and (36):

$$H_2O_{adsorbed} + h_{VB,ZnO}^{+\rightarrow \bullet OH + H_{adsorbed}^{+}} \quad (35)$$

$$h_{VB,ZnO}^{++OH^{-\rightarrow \bullet OH}} \quad (36)$$

$\bullet OH$, thus formed, has very strong oxidizing character and easily oxidizes organic pollutants, either adsorbed on the surface or very close to the surface of ZnO. During 'oxygen ionosorption', $e_{CB,ZnO}^{-}$ reacts with dissolved oxygen (O_2) to form highly oxidizing anionic superoxide radical (O_2^-) as described by equation (24). Importantly, O_2^- not only oxidizes organic pollutants but also prevents electron – hole recombination. Frequently, O_2^- gets protonated and leads formation of hydroperoxyl radical (HO_2) which subsequently producesH_2O_2 and $\bullet OH$ according to the following equations (37 - 39), referred as protonation of superoxide reaction:

$$O_2^{-+h^{+\rightleftarrows HOO\cdot/HO_2}} \quad (37)$$

$$2HOO^{\cdot} \rightarrow H_2O_2 + O_2 \quad (38)$$

$$H_2O_2 \rightarrow 2OH^{\cdot}. \quad (39)$$

In this context, it may be stated that pollutants get photocalytically degraded by either $\bullet OH$ or photogenerated charge carriers ($e^-_{CB,ZnO}$ and$h^+_{VB,ZnO}$) as described by equations (40 - 42):

Pollutant + $\bullet OH$ → CO_2 + H_2O (pollutant intermediates) (40)

Pollutants + $h^+_{VB,ZnO}$ → oxidation products (41)

Pollutants + $e^-_{CB,ZnO}$ → reduction products (42)

In general, both oxidative and reductive degradation take place on the surface of ZnO nanostructures.

(ii) Direct Degradation

In direct degradation process, electrons are excited from ground state to triplet state of the pollutant molecules in the presence of light irradiation and the excited electrons get immediately transferred into conduction band of ZnO. This charge transfer leads O_2^- generation and successively $\bullet OH$ gets formed. The reaction mechanisms can be represented by the following steps:

Pollutant + hν → Pollutant* (43)

Pollutant* + ZnO → Pollutant$^+$ + $e^-_{CB,ZnO}$ (44)

$O_2 + e_{CB,ZnO}^{-\rightarrow O_2^-}$ (45)

$\bullet OH$, thus produced, mainly contribute to oxidation. However, it has been identified that direct degradation mechanism is significant under visible radiation when pollutants absorbs visible light photon (λ > 400 nm).

3.3. General Factors Affecting the Photocatalytic Activity of ZnO

There are several factors such as pH, pollutant adsorption strength, initial concentration of pollutant, concentration of photocatalyst, addition of oxidizing species etc. to effect photocatalytic activity of ZnO. For example, pH of the solution plays important roles during (i) adsorption of pollutants on the surface of ZnO and (ii) $\bullet OH$ generation. Primarily, pH determines the surface-charge of the ZnO nanostructures. In this context, point of zero charge PZC) is defined as the pH PZC (pH_{PZC}) at which surface charge becomes zero. At pH < pH_{PZC}, ZnO surface becomes positively charged on which anionic pollutants gets adsorbed, while cationic pollutant gets adsorbed at pH > pH_{PZC}. pH also influences on the pollution degradation mechanism and rate, since lower pH helps $h^{+}_{VB,ZnO}$ mediated photocatalytic process, while $\bullet OH$ mostly contributes at higher or neutral pH (Zhang et al. 2007).

It has been investigated that competitive adsorption between water molecules and pollutant on the active sites at the surface of ZnO nanostructures plays significant role on photocatalytic activity, since oxidizing agents, generated by photo-excitation mechanism, doesn't migrate beyond a critical distance. Thus in case of pollutants far away from the surface of nanostructures, photocatalytic rate gets reduced. However, very strong attractive interaction between pollutants and nanostructures often leads formation of multiple layers on the surface of the ZnO nanostructures and direct degradation gets reduced. Hence, for efficient photocatalytic activity, choice of pollutants which will be degraded by ZnO becomes important issues. It is well-accepted that increasing intensity of light enhances degradation rate as more number of charge carriers ($h^{+}_{VB,ZnO}$ and $e^{-}_{CB,ZnO}$) are being generated by incidenting light (Rong et al. 2001, Mahmood et al. 2011). Initial concentration of ZnO, defined as the dose of the photocatalyst, has to be optimum for efficient photocatalytic activity. It has been observed that photocatalytic activity initially increases with dose, but finally it decreases. There are several factors behind this dose-dependence phenomenon such as (i) complete adsorption of the pollutant on ZnO surfaces, hence increasing ZnO doesn't increase photocatalytic activity, (ii) presence of excess ZnO increases opacity of the solution that reduces degradation rate, and (iii) higher ZnO – ZnO interaction causes deactivation of activated molecules due to collision with ZnO. Initial concentration of the pollutant has significant impact on catalytic efficiency. It has been observed that degradation rate gets decreased at higher initial concentration of pollutant due to following reasons: (i) increasing occupation of active sites by pollutant at higher pollutant concentration reduces $\bullet OH$ generation, (ii) path length of photon decreases with increasing pollutant concentration, thus light absorption by ZnO gets decreased. Sometime, it appeared that the addition of optimum oxidizing species such as H_2O_2, potassium peroxydisulfide ($K_2S_2O_8$) etc. increase the degradation rate. Mechanistically, it appears that low concentration of H_2O_2 causes charge separation which increases $\bullet OH$ concentration, but higher concentration of H_2O_2 scavenges$\bullet OH$, thus decreases catalytic degradation rate.

3.4. Different Techniques to Increase Photocatalytic Efficiency

It is well-accepted that ZnO nanostructures are efficient photocatalysts due to their high surface to volume ratio. In this context, different ZnO nanostructures such as nanowires, nanorods, nanoflower, hedge-hog etc. have been synthesized in order examine their photocatalytic activity. Variations in absorption coefficient and active sites of the respective nanostructures have been identified for observed variation in catalytic activity. In addition, ZnO facet is noticed to be another factor to influence the photocatalytic activity. For example, Zn terminated [0001] surface of ZnO nanostructure carrying positive charge preferentially attracts OH^- ions, thus $\bullet OH$ generation from [0001] surface gets increased significantly (Dodd et al. 2009). Similar to luminescence property, V_O plays a crucial role in the photocatalytic activity of ZnO, rather it exhibits multiple roles. For example, photocatalytic activity of ZnO is attributed to $h^+_{VB,ZnO}$and$e^-_{CB,ZnO}$, thus their concentration in respective bands is noticed to impact significantly on the photocatalytic degradation rate. As per previous discussion, V_O controls $h^+_{VB,ZnO}$and$e^-_{CB,ZnO}$. Hence by controlling $[V_O]$, photocatalytic degradation can be tuned (Chandrasekaran et al. 2016). E_g is observed to be reduced by V_O in some ZnO nanostructures, thus over-potential for photo-excitation of electrons is decreased and consequently photocatalytic activity gets increased (Ansari et al. 2013, Wang et al. 2012). In this context, it may be stated that V_O significantly alters photo-absorption thus influencing photocatalytic activity of ZnO (Kavitha et al. 2015). In addition, it has also been identified that V_O acts as a charge trapping centre and facilitates charge carrier separation; hence photocatalytic activity is found to be increased byV_O.

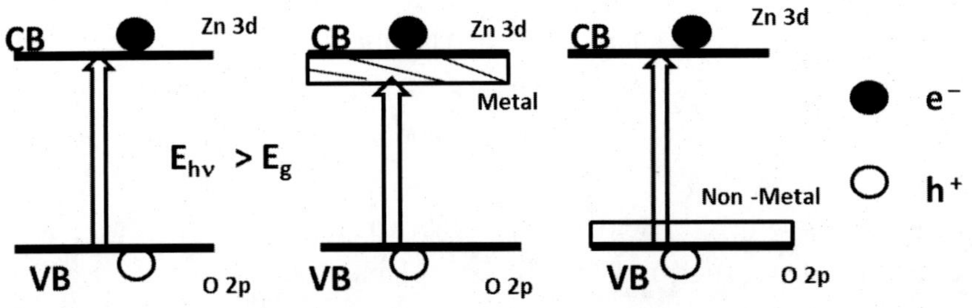

Figure 13. Schematic of the comparison of the band structures of pure, metal-doped ZnO, and nonmetal-doped ZnO.

Doping with metallic (Cu, Mn etc.) and non-metallic (N, C, S etc.) elements reducesE_g and consequently photo-absorption which primarily determines photocatalytic activity gets enhanced (Alosfur et al. 2015, Yu et al. 2016, Liu et al. 2010). In this context, it may be stated that metallic substitution alters conduction band, while non-metallic substitution modifies valence band (Shown in Figure 13).

Till date, several methods have been attempted to increase photocatalytic activity of ZnO. One such attempt is to form heterostructure of ZnO with other semiconductor or noble metal. In this context, it may be stated that the interfacial properties of ZnO nanostructures which gets modified within heterostructures enhance photocatalytic activity of ZnO. Briefly, increased carrier lifetime and interfacial charge transfer lead enhancement of photocatalytic activity of ZnO. In general, ZnO heterostructures with other semiconductors can be divided into two types depending on charge carrier separation mechanism, namely (i) conventional type – II and p-n junction and (ii) Z-scheme. In conventional type – II heterostructures, ZnO is combined with other semiconductor with proper location of conduction and valence band in such a way that conduction band of ZnO (or other semiconductor) and valence band of other semiconductor (or ZnO) take part in the interfacial charge transfer process. Thus recombination of photogenerated $h^{+}_{VB,othersemiconducto}$- $e^{-/h^{+}_{VB,ZnO}}_{CB,ZnO}$ and $e^{-}_{CB,othersemiconducto}$ is reduced and successively redox reaction generating O_2^- and $\bullet OH$ gets increased remarkably. In general, low E_g materials such as CdS, ZnSe, CuO, In_2O_3 etc. are used for this purpose (Swiegers et al. 2012). Other factors enhancing the photocatalytic activity within these heterostructures are reduced charge carrier recombination, increased active sites for O_2^- generation and visible light driven catalytic process.

p – n junction based photocatalyst is fabricated using p-type ZnO and n-type semiconductor such as TiO_2 SnO_2 etc. Mechanistically, an electric field develops at the junction due to migration of charge carriers due to difference in their respective Fermi levels in the absence of illumination. Under illumination, photogenerated charge carriers are separated by the junction in-build electric field, i.e., recombination of charge carriers gets decreased, hence catalytic activity increases significantly. It has been found that these p-n heterostructures possesses higher catalytic activity than conventional type – II heterostructures.

Z-scheme heterostructure works on completely different protocols. Briefly, two different semiconductors (one of them must be ZnO) are used to fabricate the Z-scheme heterostructure where photogenerated electrons will migrate from one semiconductor to other due to higher reduction potential while hole will still remain in the first semiconductor due to higher oxidation potential. Thus charge carrier separation gets facilitated.

Additionally, in these Z-scheme heterostructures, e_{CB}^{-} and $h_{VB}^{+}$are accumulated in semiconductor having higher reduction and oxidation potential respectively and accelerates redox reactions for photocatalytic reactions. In this context, it may be stated that ZnO/graphitic C_3N_4, ZnO/CeO_2, ZnO/graphene quantum dote graphitic heterostructures have gained attention.

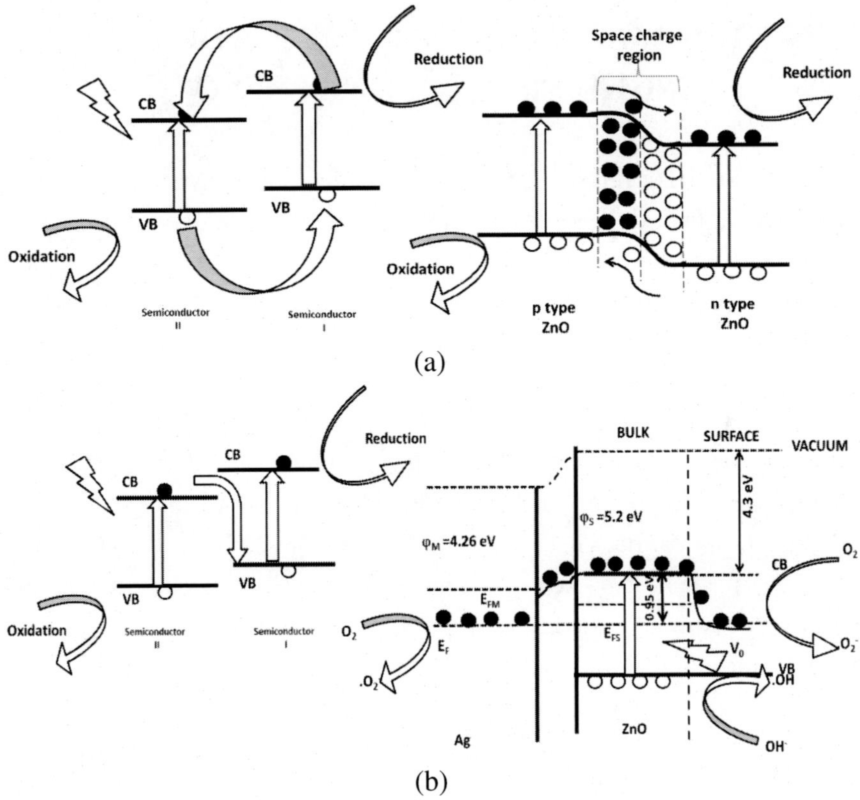

Figure 14. Heterostructure formation and electron transfer.

ZnO/novel metal heterostructure (Samanta et al. 2017) has been identified as another technique to increase photocatalytic activity. Here, novel metals such as Pt, Ag, Au etc., deposited on ZnO nanostructures, facilitates catalytic activity by increasing reactive sites as well as they act as a co-catalyst. In addition, few novel metals also increases absorption coefficient of ZnO due to surface plasmon resonance, and promotes separation of photo-generated charge carrier (as shown in Figure 14 (b)) depending on the differences in their respective work function. In this context, photocatalytic degradation of various ZnO nanostructures/ heterostructures are represented in Appendix I.

4. *Ab-initio* Understanding of the Defects and Heterostructure of ZnO

It is well-resolved from previous sections that emission in visible range and visible light induced photocatalytic activity of ZnO originate from various defect states and lots of controversies existed in this field. For understanding of the basic mechanism and reliability of the defect states, several theoretical studies have also been carried out using density functional theory (DFT) in addition to experimental investigations. Fundamentally, formation energies of the defects with reference to total energy of the perfect crystal and relevant phase are calculated here. In these calculations, local density approximation (LDA) and generalized gradient approximation (GGA) are used as exchange-correlation functionals to represent electron – electron and electron – nucleus interaction respectively during evaluation of energy. But, neither of these two exchange-correlations perfectly describes correct picture of defects of localized character. To deal localized defect states, Hubbard U correction factor is included in LDA and GGA exchange-correlation functionals. In recent time, more advanced exchange-correlation functionals such as Hartree-Fock (HF), screened exchange (sX), GW giving more realistic picture are being used for more accurate native defect structure of ZnO. Two different boundary conditions that significantly influence on energy calculation are adopted here. In one approach, defects are incorporated into a supercell which gets formed by expanding the unit cell. But this boundary condition gets limited as it has been noticed that electrostatic and elastic interactions between native defects deeply effect on the formation energy of defects. Recently, cluster approach which is free from periodic boundary interactions has been developed for more accurate calculation of native defect structure.

It has been well-studied that the parameter which represent any thermodynamic process is the Gibb's free energy and feasibility of the processes are examined by calculating change in Gibb's free energy (ΔG_f). In these calculations, ΔG_f corresponding to defect formation is primarily calculated. In this context it may be stated that ΔG_f aren't calculated directly using DFT, rather it's electronic contribution, referred as formation energy (ΔE_f), is evaluated due to following reasons: Gibb's free energy (G) consists of electronic energy (E_{el}), vibrational energy (E_{vib}), entropy from vibration (S_{vib}) according to the following expression:

$$G = E_{el} + E_{vib} - TS_{vib} + pV \tag{46}$$

where, T, p and V represent the temperature, pressure and temperature respectively. It has been noticed that E_{el} >> E_{vib}, TS_{vib}, pV for all crystalline materials at moderate temperature and pressure, hence for all practical purposes G ≈ E_{el}. In a system having very low concentration of defects, ΔG_f is expressed by equation (47):

$$\Delta G_f = G^{defect} - \sum_i N_i \mu_i + qE_F \tag{47}$$

Where, G^{defect}, μ_i and N_i represent Gibb's free energy of the system with a defect state in 'q' charge state, number and chemical potential of ith constituent atom respectively. With the approximation that $G_f \approx E_{el}$, above relation gets simplified into

$$G_f = E_{el}^{defect} - \sum_i N_i \mu_i + qE_F \tag{48}$$

Here, μ_i and E_F are evaluated with respect to a standard chemical potential (μ_i^0) and energy of the valence band maxima (E_{VBM}) and the following expressions for μ_i and E_F are substituted in equation (49):

$$\Delta\mu_i = \mu_i - \mu_i^0 \text{ and } \Delta E_F = E_F - E_{VBM} \tag{49}$$

Thus ΔE_f gets simplified into the following relation:

$$\Delta E_f = E_{el}^{defect} - \sum_i N_i\left(\Delta\mu_i + \mu_i^0\right) + q(\Delta E_F + E_{VBM}) \tag{50}$$

i.e.,

$$\Delta E_f = E_{el}^{defect} - N_{Zn}(\Delta\mu_{Zn} + \mu_{Zn}^0) - N_O\left(\Delta\mu_O + \mu_O^0\right) + q(\Delta E_F + E_{VBM}) \tag{51}$$

If ΔN_{Zn} and ΔN_O denote the difference in number of Zn and O atoms with respect to perfect crystal, the above equation can be written in the following form:

$$\Delta E_f = E_{el}^{defect} - E_{el}^{perfect} - \Delta N_{Zn}(\Delta\mu_{Zn} + E_{Zn}) - \Delta N_O(\Delta\mu_O + E_O) + q(\Delta E_F + E_{VBM}) \tag{52}$$

Here, $E_{el}^{perfect}$ represents the energy of the perfect supercell. E_{Zn} and E_O are the total energy of Zn crystal and O_2 molecule per atom. Fundamentals of defect mechanism and their reliability corresponding to all defect states can be understood from equation (51). For example, ΔN_{Zn} = 0 and ΔN_O = -1 represent V_O, while ΔN_{Zn} = 1 and ΔN_O = 0 denote Zn_i. For defects with variable charge state, the feasibility of defect formation is achieved from the condition at which formation energies of different defect states become equal. It has been found that formation energies of the oxygen vacancies of various charge states (V_O, V_O^+ and V_O^{++}) significantly depend on the location of Fermi energy. At lower Fermi energy, V_O^{++} is found to be most stable, while at higher Fermi energy V_O is favourable while the calculation excludes V_O^+formation. Thermodynamic transition level for $V_O \rightarrow V_O^{++}$ is calculated to be 2.2 eV above valence band (1.2 eV below conduction band), i.e., V_O creates a deep level donor state. ForZn_i defect states, octahedral sites are most favourable, while its occupancy at tetrahedral site is comparatively less favourable. It's thermodynamic transition levels $\epsilon'(2++0)$ lie within 0.05 eV of conduction band minima. Thus Zn_i creates shallow donor levels (Look et al. 1999). In this context, it may be stated that the formation energy of Zn_i under O-deficient condition is considerably very high (~ 4 eV), thus the probability of Zn_i formation appears to be very low under this condition, but reverse phenomenon happens under O-rich condition as Zn_i has low formation energy ~ 0.5 eV. Zinc vacancy (V_{Zn}) was supposed to be an acceptor-type defect responsible for n-type conductivity (Janotti et al. 2007). However, recent theoretical studies illustrate that V_{Zn} exhibits two deep acceptor levels ϵ' and $\epsilon'(--2)$ at 0.7 and 2.4 eV respectively and doesn't have contribution to hole generation, but it influences significantly on green luminescence. O_i is suggested to exists wither in octahedral interstitial site or as O_2-molecule-like configuration. Former configuration has higher formation energy in comparison to V_{Zn} and exhibits deep acceptor levels ϵ' and $\epsilon'(--2)$ at 0.72 and 1.59 eV respectively. Later possesses even higher formation energy. However, it has been noticed that charge configuration remains unchanged over whole range of Fermi energy, thus O_i is suggested to be electrically inactive. Similar results are also obtained for O-anti-site in which O occupies Zn sites. When O sites are occupied by Zn ions, it is referred as Zn-antisite. Zn-antisite shows transition level of $\epsilon'(2++0)$ near conduction band minima, similar to Zn_i and two deep (ϵ' andϵ') states at the middle of E_g. It has been calculated that the formation energy of Zn-antisite is very high under n-type condition, whereas it is very low under p-type condition. Therefore, it may be safely stated that most controversial green emission originate from V_O^{++} while V_{Zn} significantly influences lifetime, intensity, FWHM of the emission.

APPENDIX I

Table 1. Photocatalytic Activity of Different ZnO Nanostructures

Sl. No.	Nanocomposite	UV/ Visible light activated	Rate constant	Used Dye	Reference
1	ZnO nanorods	Visible-light	0.0215 min^{-1}	MO	Tian, C., Zhang, Q., Wu, A., Jiang, M., Liang, Z., Jiang, B., & Fu, H. (2012). Cost-effective large-scale synthesis of ZnO photocatalyst with excellent performance for dye photodegradation. *Chemical Communications*, *48*(23), 2858-2860
2	Pit-structured ZnO nanorods	UV light	0.01402 min^{-1}	MB	Wu, D., Wang, W., Tan, F., Sun, F., Lu, H., & Qiao, X. (2013). Fabrication of pit-structured ZnO nanorods and their enhanced photocatalytic performance. *RSC Advances*, *3*(43), 20054-20059
3	ZnO nanorods	UV	0.036 min^{-1}	Orange G	Leelavathi, a Giridhar Madrasa and N. Ravishankar*b Origin of enhanced photocatalytic activity and photoconduction in high aspect ratio ZnO nanorods† *Phys. Chem. Chem. Phys.*, 2013, 15, 10795—10802
4	ZnO nanorods	UV-light	0.0134 min^{-1}	MB	Ma, L., Yang, X., Zhou, Z. Q., & Lu, M. (2016). A synergetic effect of surface plasmon and ammoniation on enhancing photocatalytic activity of ZnO nanorods. *RSC Advances*, *6*(100), 97808-97817.
5	ZnO nanostructures	Sunlight UV-light	0.075min^{-1} 0.037 min^{-1}	MB	Hezam, A., Namratha, K., Drmosh, Q. A., Chandrashekar, B. N., Sadasivuni, K. K., Yamani, Z. H., ... & Byrappa, K. (2017). Heterogeneous growth mechanism of ZnO nanostructures and the effects of their morphology on optical and photocatalytic properties. *CrystEngComm*, *19*(24), 3299-3312.
6	Nano needle decorated ZnO hollow spheres	UV-light	0.0732 min^{-1}	RhB	Liu, Y., Liu, H., Huang, B., Li, T., & Wang, J. (2017). Nano needle decorated ZnO hollow spheres with exposed (0001) planes and their corrosion using acetic acid. CrystEngComm, 19(38), 5774-5779

Sl. No.	Nanocomposite	UV/ Visible light activated	Rate constant	Used Dye	Reference
7	ZnO nanowire arrays	UV-VIS	0.0225 min^{-1}	MB	Dao, T. D., Han, G., Arai, N., Nabatame, T., Wada, Y., Hoang, C. V., ... & Nagao, T. (2015). Plasmon-mediated photocatalytic activity of wet-chemically prepared ZnO nanowire arrays. *Physical Chemistry Chemical Physics*, *17*(11), 7395-7403
8	flower-like ZnO hierarchical microspheres	Hg lamp	0.0321 min^{-1}	Rhodamine 6G	Jinxin Zhan, a Hongxing Dong,*a Yang Liu, a Yinglei Wang, b Zhanghai Chenb and Long Zhang*a A novel synthesis and excellent photodegradation of flower-like ZnO hierarchical microspheres CrystEngComm, 2013, 15, 10272–10277
9	ZnO nanomaterials	sunlight	0.022min^{-1}	tetracycline	Wang, Q., Zhou, H., Liu, X., Li, T., Jiang, C., Song, W., & Chen, W. (2018). Facet-dependent generation of superoxide radical anions by ZnO nanomaterials under simulated solar light. *Environmental Science: Nano*, *5*(12), 2864-2875.
10	ZnO nanorods Nanofowers Nanopyramids nanoprisms	UV light	0.0632 min^{-1} 0.0291 min^{-1} 0.0153 min^{-1} 0.0105 min^{-1}	MR	Ahmed, F., Arshi, N., Anwar, M. S., Danish, R., & Koo, B. H. (2014). Morphological evolution of ZnO nanostructures and their aspect ratio-induced enhancement in photocatalytic properties. *RSC Advances*, *4*(55), 29249-29263
11	zinc oxide nanocrystals	UV light	0.085 min^{-1}	Methylene blue	Prakash, A., & Bahadur, D. (2014). Chemically derived defects in zinc oxide nanocrystals and their enhanced photo-electrocatalytic activities. *Physical Chemistry Chemical Physics*, *16*(39), 21429-21437
12	Pure ZnO	Visible-light	0.089 h^{-1} 0.079 h^{-1} 0.086 h^{-1}	4-nitrophenol methylene blue brilliant blue	Ansari, S. A., Khan, M. M., Kalathil, S., Nisar, A., Lee, J., & Cho, M. H. (2013). Oxygen vacancy induced band gap narrowing of ZnO nanostructures by an electrochemically active biofilm. *Nanoscale*, *5*(19), 9238-9246
13	Facile fabrication of highly efficient modified ZnO photocatalyst	Visible-light	0.101 min^{-1}	MB	Raza, W., Faisal, S. M., Owais, M., Bahnemann, D., & Muneer, M. (2016). Facile fabrication of highly efficient modified ZnO photocatalyst with enhanced photocatalytic, antibacterial and anticancer activity. *RSC Advances*, *6*(82), 78335-78350.

Table 1. (Continued)

Sl. No.	Nanocomposite	UV/ Visible light activated	Rate constant	Used Dye	Reference
14	ZnO_{1-x}	Visible-light	0.522 h^{-1}	MB	Lv, Y., Yao, W., Ma, X., Pan, C., Zong, R., & Zhu, Y. (2013). The surface oxygen vacancy induced visible activity and enhanced UV activity of a ZnO 1− x photocatalyst. *Catalysis Science & Technology*, *3*(12), 3136-3146
15	Spongy-ZnO	Visible & Solar light	0.035 min^{-1}	Crystal violet (CV)	Adhikari, S., Gupta, R., Surin, A., Kumar, T. S., Chakraborty, S., Sarkar, D., & Madras, G. (2016). Visible light assisted improved photocatalytic activity of combustion synthesized spongy-ZnO towards dye degradation and bacterial inactivation. RSC Advances, 6(83), 80086-80098
16	Hierarchical ZnO nanostructures	UV light	2.30 min^{-1} 0.022 min^{-1}	MB Rhodamine B	Singh, S., Barick, K. C., & Bahadur, D. (2013). Shape-controlled hierarchical ZnO architectures: photocatalytic and antibacterial activities. *CrystEngComm*, *15*(23), 4631-4639
17	ZnO meso- /nanocrystals	UV-light	0.020 min^{-1}	MO	Liang, S., Gou, X., Cui, J., Luo, Y., Qu, H., Zhang, T., ... & Sun, S. (2018). Novel cone-like ZnO mesocrystals with co-exposed (101 [combining macron] 1) and (0001 [combining macron]) facets and enhanced photocatalytic activity. *Inorganic Chemistry Frontiers*, *5*(9), 2257-2267.
18	ZnO inverse opal film	UV-Vis light (250 nm–780 nm)	0.0137 min^{-1}	Rhodamine B	Liu, J., Jin, J., Li, Y., Huang, H. W., Wang, C., Wu, M., ... & Su, B. L. (2014). Tracing the slow photon effect in a ZnO inverse opal film for photocatalytic activity enhancement. *Journal of Materials Chemistry A*, *2*(14), 5051-5059
19	C-doped ZnO nanosheet	Visible-light	0.006 min^{-1}	Rhodamine B	Liu, S., Li, C., Yu, J., & Xiang, Q. (2011). Improved visible-light photocatalytic activity of porous carbon self-doped ZnO nanosheet-assembled flowers. *CrystEngComm*, *13*(7), 2533-2541

Sl. No.	Nanocomposite	UV/ Visible light activated	Rate constant	Used Dye	Reference
20	ZnO/C quantum dot nanocomposite	Visible-light	0.095 h^{-1}	Gas phase benzene	Yan, X., Zou, C., Gao, X., & Gao, W. (2012). ZnO/TiO 2 core–brush nanostructure: processing, microstructure and enhanced photocatalytic activity. *Journal of Materials Chemistry*, *22*(12), 5629-5640
21	ZnO/C composites	UV light	0.326 min^{-1}	Methylene blue	Ma, S., Xue, J., Zhou, Y., Zhang, Z., & Wu, X. (2014). A facile route for the preparation of ZnO/C composites with high photocatalytic activity and adsorption capacity. *CrystEngComm*, *16*(21), 4478-4484
22	carbon-doped ZnO nanostructures	Visible-light	0.525 h^{-1}	RhB	Ansari, S. A., Ansari, S. G., Foaud, H., & Cho, M. H. (2017). Facile and sustainable synthesis of carbon-doped ZnO nanostructures towards the superior visible light photocatalytic performance. *New Journal of Chemistry*, *41*(17), 9314-9320.
23	Iodine doped ZnO nanoflower	Visible-light	0.0074 min^{-1}	Rhodamine B	Barka-Bouaifel, F., Sieber, B., Bezzi, N., Benner, J., Roussel, P., Boussekey, L., ... & Boukherroub, R. (2011). Synthesis and photocatalytic activity of iodine-doped ZnO nanoflowers. *Journal of Materials Chemistry*, *21*(29), 10982-10989.
24	Na-doped p-type flower-like ZnO	Visible-light	0.0176 min^{-1}	MB	Kim, K. J., Kreider, P. B., Choi, C., Chang, C. H., & Ahn, H. G. (2013). Visible-light-sensitive Na-doped p-type flower-like ZnO photocatalysts synthesized via a continuous flow microreactor. *RSC Advances*, *3*(31), 12702-12710
25	N-doped ZnO with cabbage morphology	UV-Vis-light	0.0585 min^{-1}	MB	Prabakaran, E., & Pillay, K. (2019). Synthesis of N-doped ZnO nanoparticles with cabbage morphology as a catalyst for the efficient photocatalytic degradation of methylene blue under UV and visible light. RSC Advances, 9(13), 7509-7535
26	ZnO/Au heterostructure	UV-light	0.0622 min^{-1}	MO	Xiao, F., Wang, F., Fu, X., & Zheng, Y. (2012). A green and facile self-assembly preparation of gold nanoparticles/ZnO nanocomposite for photocatalytic and photoelectrochemical applications. *Journal of Materials Chemistry*, *22*(7), 2868-2877

Table 1. (Continued)

Sl. No.	Nanocomposite	UV/ Visible light activated	Rate constant	Used Dye	Reference
27	ZnO/Au and ZnO/Ag nanoparticles	Visible-light	0.08 min^{-1} 0.004 min^{-1}	MB	Fageria, P., Gangopadhyay, S., & Pande, S. (2014). Synthesis of ZnO/Au and ZnO/Ag nanoparticles and their photocatalytic application using UV and visible light. *Rsc Advances*, *4*(48), 24962-24972
28	Au@ZnO core–shell nanostructures	UV-light	0.042 min^{-1}	RhB	Shao, X., Li, B., Zhang, B., Shao, L., & Wu, Y. (2016). Au@ ZnO core–shell nanostructures with plasmon-induced visible-light photocatalytic and photoelectrochemical properties. *Inorganic Chemistry Frontiers*, *3*(7), 934-943.
29	Au nanoparticles in situ grown on ZnO nanorods	UV-light	0.332 min^{-1}	RhB	Song, X., Zhang, X., & Yang, P. (2016). In situ growth of small Au nanoparticles on ZnO nanorods via ultrasonic irradiation toward super-enhanced catalytic activity. *RSC Advances*, *6*(109), 107433-107441.
30	Au–ZnO nanopyramids	Visible	0.059 min^{-1}	MB	Ranasingha, O. K., Wang, C., Ohodnicki, P. R., Lekse, J. W., Lewis, J. P., & Matranga, C. (2015). Synthesis, characterization, and photocatalytic activity of Au–ZnO nanopyramids. Journal of Materials Chemistry A, 3(29), 15141-15147
31	Au@ZnO–Pd ternary core–shell nanostructures	UV-light	0.058 min^{-1} And 0.25 min^{-1}	PHENOL and MB	Li, B., Wang, R., Shao, X., Shao, L., & Zhang, B. (2017). Synergistically enhanced photocatalysis from plasmonics and a co-catalyst in Au@ ZnO–Pd ternary core–shell nanostructures. *Inorganic Chemistry Frontiers*, *4*(12), 2088-2096.
32	Ag nanoparticle/ ZnO Nanorod heterostructures	UV light		Rhodamine B	Wu, Z., Xu, C., Wu, Y., Yu, H., Tao, Y., Wan, H., & Gao, F. (2013). ZnO nanorods/Ag nanoparticles heterostructures with tunable Ag contents: A facile solution-phase synthesis and applications in photocatalysis. *CrystEngComm*, *15*(30), 5994-6002

Sl. No.	Nanocomposite	UV/ Visible light activated	Rate constant	Used Dye	Reference
33	Ag–ZnO hybrid nanospindles	Sunlight	0.126 min^{-1}	Methyl orange	Kuriakose, S., Choudhary, V., Satpati, B., & Mohapatra, S. (2014). Facile synthesis of Ag–ZnO hybrid nanospindles for highly efficient photocatalytic degradation of methyl orange. *Physical Chemistry Chemical Physics*, *16*(33), 17560-17568
34	ZnO@Ag core–shell nanorods	Solar-light	0.03239 min^{-1} 0.01867 min^{-1} 0.01474 min^{-1}	Rhodamine 6G Congo red Amido Black	Dinesh, V. P., Biji, P., Ashok, A., Dhara, S. K., Kamruddin, M., Tyagi, A. K., & Raj, B. (2014). Plasmon-mediated, highly enhanced photocatalytic degradation of industrial textile dyes using hybrid ZnO@ Ag core–shell nanorods. *RSC Advances*, *4*(103), 58930-58940
35	Ag/ZnO prismatic nanorods	UV-light	0.025min^{-1}	RB	Xin, Z., Li, L., Zhang, X., & Zhang, W. (2018). Microwave-assisted hydrothermal synthesis of chrysanthemum-like Ag/ZnO prismatic nanorods and their photocatalytic properties with multiple modes for dye degradation and hydrogen production. *RSC advances*, *8*(11), 6027-6038.
36	Ag–ZnO nano particle	sunlight	0.002 min^{-1}	Rhodamine B	Choudhary, M. K., Kataria, J., Bhardwaj, V. K., & Sharma, S. (2019). Green biomimetic preparation of efficient Ag–ZnO heterojunctions with excellent photocatalytic performance under solar light irradiation: a novel biogenic-deposition-precipitation approach. *Nanoscale Advances*, *1*(3), 1035-1044.
37	ZnO/Ag micro/nanospheres	mercury lamp	0.1189 min^{-1}	MO	Li, Z., Zhang, F., Meng, A., Xie, C., & Xing, J. (2015). ZnO/Ag micro/nanospheres with enhanced photocatalytic and antibacterial properties synthesized by a novel continuous synthesis method. *RSC Advances*, *5*(1), 612-620.
38	ZnO/Ag heterostructures	Visible	0.0716 min^{-1}	Xylenol orange (XO)	Dong, Y., Feng, C., Jiang, P., Wang, G., Li, K., & Miao, H. (2014). Simple one-pot synthesis of ZnO/Ag heterostructures and the application in visible-light-responsive photocatalysis. *RSC Advances*, *4*(14), 7340-7346.

Table 1. (Continued)

Sl. No.	Nanocomposite	UV/ Visible light activated	Rate constant	Used Dye	Reference
39	Ag/ZnO metal–semiconductor	UV-light	0.023 min^{-1}	MB	Shaislamov, U., & Lee, H. J. (2018). Facile synthesis of Ag/ZnO metal–semiconductor hierarchical photocatalyst nanostructures via the galvanic-potential-enhanced hydrothermal method. CrystEngComm, 20(46), 7492-7501
40	Ag–ZnO nanoparticles	Visible-light	0.006 min^{-1}	MO	Tao, S., Yang, M., Chen, H., Ren, M., & Chen, G. (2016). Continuous synthesis of hedgehog-like Ag–ZnO nanoparticles in a two-stage microfluidic system. *RSC Advances*, *6*(51), 45503-45511.
41	ZnO/TiO2 core–brush nanostructure	UV-light (245 nm) Visible-light (450 nm)	0.764% min^{-1} 0.247% min^{-1}	Br-PGR	Yan, X., Zou, C., Gao, X., & Gao, W. (2012). ZnO/TiO 2 core–brush nanostructure: processing, microstructure and enhanced photocatalytic activity. Journal of Materials Chemistry, 22(12), 5629-5640
42	ZnO/TiO2 vertical-nanoneedle-onfilm heterojunction	mercury lamp	0.3775 h^{-1}	methylene blue	Li, D., Zhang, Y., Wu, W., & Pan, C. (2014). Preparation of a ZnO/TiO 2 vertical-nanoneedle-on-film heterojunction and its photocatalytic properties. *RSC Advances*, *4*(35), 18186-18192
43	ZnO@TiO_2 nanotubes heterostructured	UV-light	0.034min^{-1}	MO	Liao, Y., Zhang, K., Wang, X., Zhang, D., Li, Y., Su, H., ... & Zhong, Z. (2018). Preparation of ZnO@ TiO 2 nanotubes heterostructured film by thermal decomposition and their photocatalytic performances. *RSC advances*, *8*(15), 8064-8070.
44	anatase–TiO_2 (B) biphase nanowire/ZnO nanoparticle heterojunction	UV-light	0.0420 min^{-1}	MO	Sun, C., Xu, Q., Xie, Y., Ling, Y., & Hou, Y. (2018). Designed synthesis of anatase–TiO_2 (B) biphase nanowire/ZnO nanoparticle heterojunction for enhanced photocatalysis. *Journal of Materials Chemistry A*, *6*(18), 8289-8298.

Sl. No.	Nanocomposite	UV/ Visible light activated	Rate constant	Used Dye	Reference
45	Electrospun TiO_2/ZnO/PAN hybrid nanofiber	UV-light	0.014 min^{-1}	MG	Yar, A., Haspulat, B., Üstün, T., Eskizeybek, V., Avcı, A., Kamış, H., & Achour, S. (2017). Electrospun TiO 2/ZnO/PAN hybrid nanofiber membranes with efficient photocatalytic activity. *RSC advances*, *7*(47), 29806-29814.
46	Mn/ZnO Mn-doped ZnO nanorods	Visible-light	0.301 h^{-1}	2,4-dichlorophenol (DCP)	Lu, Y., Lin, Y., Xie, T., Shi, S., Fan, H., & Wang, D. (2012). Enhancement of visible-light-driven photoresponse of Mn/ZnO system: photogenerated charge transfer properties and photocatalytic activity. Nanoscale, 4(20), 6393-6400
47	ZnO nanorods/Pt heteronanostructure	UV light	0.169 min^{-1}	Rhodamine B	Wu, Z., Xue, Y., Wang, H., Wu, Y., & Yu, H. (2014). ZnO nanorods/Pt and ZnO nanorods/Ag heteronanostructure arrays with enhanced photocatalytic degradation of dyes. *RSC Advances*, *4*(103), 59009-59016
48	Pt-ZnO nanocomposite microspheres	UV light	0.3465 h^{-1}	Orange-II (4-(2-hydroxy-1-givnaphthylazo)Benzenesulfonic acid)	Yu, C., Yang, K., Xie, Y., Fan, Q., Jimmy, C. Y., Shu, Q., & Wang, C. (2013). Novel hollow Pt-ZnO nanocomposite microspheres with hierarchical structure and enhanced photocatalytic activity and stability. *Nanoscale*, *5*(5), 2142-2151
49	Pt nanoparticle loaded ZnO nanoflakes	Xe arc lamp (300 W)	0.0262 min^{-1}	MO	Ong, W. L., Natarajan, S., Kloostra, B., & Ho, G. W. (2013). Metal nanoparticle-loaded hierarchically assembled ZnO nanoflakes for enhanced photocatalytic performance. *Nanoscale*, *5*(12), 5568-5575
50	ZnO/graphite C_3N_4 heterostructure	Visible-light	0.0305 min^{-1}	MB	Wang, Y., Shi, R., Lin, J., & Zhu, Y. (2011). Enhancement of photocurrent and photocatalytic activity of ZnO hybridized with graphite-like C_3N_4. *Energy & Environmental Science*, *4*(8), 2922-2929
51	ZnO/graphite C_3N_4 heterostructure	Visible-light	0.035 min^{-1}	MO	Sun, J. X., Yuan, Y. P., Qiu, L. G., Jiang, X., Xie, A. J., Shen, Y. H., & Zhu, J. F. (2012). Fabrication of composite photocatalyst gC_3N_4–ZnO and enhancement of photocatalytic activity under visible light. *Dalton Transactions*, *41*(22), 6756-6763.
52	N-doped ZnO/g-C_3N_4 core–shell nanoplates	Visible-light	0.0441 min^{-1}	Rhodamine B	Kumar, S., Baruah, A., Tonda, S., Kumar, B., Shanker, V., & Sreedhar, B. (2014). Cost-effective and eco-friendly synthesis of novel and stable N-doped ZnO/gC_3N_4 core–shell nanoplates with excellent visible-light responsive photocatalysis. *Nanoscale*, *6*(9), 4830-4842

Table 1. (Continued)

Sl. No.	Nanocomposite	UV/ Visible light activated	Rate constant	Used Dye	Reference
53	ZnO/mpg-C_3N_4 heterojunction	Visible-light	0.182 h^{-1}	Methylene blue	Chen, D., Wang, K., Ren, T., Ding, H., & Zhu, Y. (2014). Synthesis and characterization of the ZnO/mpg-C_3N_4 heterojunction photocatalyst with enhanced visible light photoactivity. *Dalton Transactions*, *43*(34), 13105-13114
54	Ultra-thin coating of g-C3N4 on an aligned ZnO nanorod	Visible light	0.044 min^{-1}	MB	Park, T. J., Pawar, R. C., Kang, S., & Lee, C. S. (2016). Ultra-thin coating of gC 3 N 4 on an aligned ZnO nanorod film for rapid charge separation and improved photodegradation performance. *RSC Advances*, *6*(92), 89944-89952.
55	ZnO crystals/ graphene oxide	Visible-light	0.0028 min^{-1}	Cr(VI)	Pan, X., Yang, M. Q., & Xu, Y. J. (2014). Morphology control, defect engineering and photoactivity tuning of ZnO crystals by graphene oxide–a unique 2D macromolecular surfactant. *Physical Chemistry Chemical Physics*, *16*(12), 5589-5599.
56	ZnO nanorods on reduced graphene sheets	UV light	0.0268 min^{-1}	MB	Zou, R., He, G., Xu, K., Liu, Q., Zhang, Z., & Hu, J. (2013). ZnO nanorods on reduced graphene sheets with excellent field emission, gas sensor and photocatalytic properties. *Journal of Materials Chemistry A*, *1*(29), 8445-8452
57	ZnO nanorod (NR)-reduced graphene oxide (rGO) nanocomposites	UV-light	0.089 min^{-1} 0.028 min^{-1}	RhB and PHENOL	Wang, F., Zhou, Y., Pan, X., Lu, B., Huang, J., & Ye, Z. (2018). Enhanced photocatalytic properties of ZnO nanorods by electrostatic self-assembly with reduced graphene oxide. *Physical Chemistry Chemical Physics*, *20*(10), 6959-6969.
58	ZnO:Sr nanoparticles@Graphene nanospreads	Visible light	0.015 min^{-1}	MB	Ravichandran, K., Chidhambaram, N., Arun, T., Velmathi, S., & Gobalakrishnan, S. (2016). Realizing cost-effective ZnO: Sr nanoparticles@ graphene nanospreads for improved photocatalytic and antibacterial activities. *RSC Advances*, *6*(72), 67575-67585.

Sl. No.	Nanocomposite	UV/ Visible light activated	Rate constant	Used Dye	Reference
59	Porous ZnO/ZnSe nanocomposite	Visible-light	0.00523 min^{-1}	Orange-II (4-(2-hydroxy-1-naphthylazo)Benzenesulfonic acid)	Cho, S., Jang, J. W., Lee, J. S., & Lee, K. H. (2012). Porous ZnO–ZnSe nanocomposites for visible light photocatalysis. *Nanoscale*, *4*(6), 2066-2071
60	NiO–ZnO composite hollow microspheres	UV light	2.456 h^{-1} 3.99 h^{-1} 3.13 h^{-1}	RhB MB MO	Xie, Q., Guo, H., Zhang, X., Lu, A., Zeng, D., Chen, Y., & Peng, D. L. (2013). A facile approach to fabrication of well-dispersed NiO–ZnO composite hollow microspheres. *RSC Advances*, *3*(46), 24430-24439
61	Fe_3O_4 embedded ZnO nanocomposites	UV light	0.015 min^{-1}	organic dyes	Singh, S., Barick, K. C., & Bahadur, D. (2013). Fe_3O_4 embedded ZnO nanocomposites for the removal of toxic metal ions, organic dyes and bacterial pathogens. *Journal of Materials Chemistry A*, *1*(10), 3325-3333
62	$Fe_2O_3/ZnFe_2O_4$ and $ZnO/ZnFe_2O_4$ hollow nanospheres	Visible	0.0074 min^{-1}	RhB	Li, J., Liu, Z., & Zhu, Z. (2014). Magnetically separable ZnFe2O4, $Fe_2O_3/ZnFe_2O_4$ and $ZnO/ZnFe_2O_4$ hollow nanospheres with enhanced visible photocatalytic properties. RSC Advances, 4(93), 51302-51308
63	Wurtzite ZnS nanosheets/ porous ZnO nanostructures	Visible-light	0.171 min^{-1}	MB	Kole, A. K., Tiwary, C. S., & Kumbhakar, P. (2013). Ethylenediamine assisted synthesis of wurtzite zinc sulphide nanosheets and porous zinc oxide nanostructures: near white light photoluminescence emission and photocatalytic activity under visible light irradiation. *CrystEngComm*, *15*(27), 5515-5525
64	ZnS/ZnO nanosheets	UV-light	0.0489 min^{-1}	RhB	Li, X., Li, X., Zhu, B., Wang, J., Lan, H., & Chen, X. (2017). Synthesis of porous ZnS, ZnO and ZnS/ZnO nanosheets and their photocatalytic properties. *RSC Advances*, *7*(49), 30956-30962.
65	n-/p-type ZnO nanorods by lithium substitution	UV light	0.112 min^{-1}	Rhodamine B	Thakur, I., Chatterjee, S., Swain, S., Ghosh, A., Behera, S. K., & Chaudhary, Y. S. (2015). Facile synthesis of single crystalline n-/p-type ZnO nanorods by lithium substitution and their photoluminescence, electrochemical and photocatalytic properties. *New Journal of Chemistry*, *39*(4), 2612-2619

Table 1. (Continued)

Sl. No.	Nanocomposite	UV/ Visible light activated	Rate constant	Used Dye	Reference
66	Cu_2O/ZnO hetero-nanorod	Visible-light	0.0077 min^{-1}	MO	Zou, X., Fan, H., Tian, Y., & Yan, S. (2014). Synthesis of Cu 2 O/ZnO hetero-nanorod arrays with enhanced visible light-driven photocatalytic activity. *CrystEngComm*, *16*(6), 1149-1156
67	ZnO/CuS heterostructure	Visible-light	0.023 min^{-1}	Methylene blue	Basu, M., Garg, N., & Ganguli, A. K. (2014). A type-II semiconductor (ZnO/CuS heterostructure) for visible light photocatalysis. *Journal of Materials Chemistry A*, *2*(20), 7517-7525
68	ZnO@C-6 Ag decorated Coreshell structure	Visible-light	0.0302 min^{-1}	RhB	Sawant, S. Y., Kim, J. Y., Han, T. H., Ansari, S. A., & Cho, M. H. (2018). Electrochemically active biofilm-assisted biogenic synthesis of an Ag-decorated ZnO@ C core–shell ternary plasmonic photocatalyst with enhanced visible-photocatalytic activity. *New Journal of Chemistry*, *42*(3), 1995-2005.
69	In_2O_3/ZnO@Ag nanowire	Visible light	0.0099 min^{-1}	MO	Liu, H., Hu, C., Zhai, H., Yang, J., Liu, X., & Jia, H. (2017). Fabrication of In 2 O 3/ZnO@ Ag nanowire ternary composites with enhanced visible light photocatalytic activity. *RSC Advances*, *7*(59), 37220-37229.
70	Ag_2O/tetrapod-ZnO nanostructures	UV-light	0.212 min^{-1}	MB	Sun, C., Fu, Y., Wang, Q., Xing, L., Liu, B., & Xue, X. (2016). Ultrafast piezo-photocatalytic degradation of organic pollutions by Ag 2 O/tetrapod-ZnO nanostructures under ultrasonic/UV exposure. *RSC Advances*, *6*(90), 87446-87453.
71	ZnO/Ag_2O heterostructures	UV-light	1.34 min^{-1}	MB	Ma, S., Xue, J., Zhou, Y., & Zhang, Z. (2014). Photochemical synthesis of ZnO/Ag 2 O heterostructures with enhanced ultraviolet and visible photocatalytic activity. *Journal of Materials Chemistry A*, *2*(20), 7272-7280
72	Ag_2S@ZnO hybrid	Visible light	.002min^{-1}	MB	Zhang, Y., Liu, C., Zhu, G., Huang, X., Liu, W., Hu, W., ... & Zhai, J. (2017). Piezotronic-effect-enhanced Ag 2 S/ZnO photocatalyst for organic dye degradation. *RSC Advances*, *7*(76), 48176-48183.

Sl. No.	Nanocomposite	UV/ Visible light activated	Rate constant	Used Dye	Reference
73	ZnO/SnO_2 porous nanofibers	mercury lamp	0.102337 min^{-1}	methylene blue	Chen, X., Zhang, F., Wang, Q., Han, X., Li, X., Liu, J., ... & Qu, F. (2015). The synthesis of ZnO/SnO $_2$ porous nanofibers for dye adsorption and degradation. *Dalton Transactions*, *44*(7), 3034-3042
74	ZnO/SnO_2	UV-light	0.002 min^{-1}	MB	Marković, S., Stanković, A., Dostanić, J., Veselinović, L., Mančić, L., Škapin, S. D., ... & Uskoković, D. (2017). Simultaneous enhancement of natural sunlight-and artificial UV-driven photocatalytic activity of a mechanically activated ZnO/SnO 2 composite. *RSC Advances*, *7*(68), 42725-42737.
75	Type-II ZnO nanorod–SnO_2 nanoparticle heterostructures	UV light	0.017 min^{-1}	Rhodamine B	Huang, X., Shang, L., Chen, S., Xia, J., Qi, X., Wang, X., ... & Meng, X. M. (2013). Type-II ZnO nanorod–SnO $_2$ nanoparticle heterostructures: characterization of structural, optical and photocatalytic properties. *Nanoscale*, *5*(9), 3828-3833
76	3D porous ZnO–SnS p–n heterojunction	Visible-light	0.05 min^{-1}	RhB	Wang, L., Zhai, H., Jin, G., Li, X., Dong, C., Zhang, H., ... & Sun, H. (2017). 3D porous ZnO–SnS p–n heterojunction for visible light driven photocatalysis. Physical Chemistry Chemical Physics, 19(25), 16576-16585
77	SnS/ZnO nanoparticle	sunlight	0.0245 min^{-1}	MO	Jayswal, S., & Moirangthem, R. S. (2018). Construction of a solar spectrum active SnS/ZnO p–n heterojunction as a highly efficient photocatalyst: the effect of the sensitization process on its performance. *New Journal of Chemistry*, *42*(16), 13689-13701.
78	$ZnO/ZnFe_2O_4$ nanocomposite	UV-light Visible-light NIR-light	0.008 min^{-1} 0.001 min^{-1} 0.002 min^{-1}	MO presence of H_2O_2	ZnO/ZnFe2O4 nanocomposite as a broadspectrum photo-Fenton-like photocatalyst with near-infrared activity Huabi
79	Bi_2S_3–ZnO heterostructure	UV-A light	0.018 min^{-1}	Acid Black 1	Balachandran, S., & Swaminathan, M. (2013). The simple, template free synthesis of a Bi 2 S 3–ZnO heterostructure and its superior photocatalytic activity under UV-A light. *Dalton Transactions*, *42*(15), 5338-5347

Table 1. (Continued)

Sl. No.	Nanocomposite	UV/ Visible light activated	Rate constant	Used Dye	Reference
80	$Bi_2Sn_2O_7$-ZnO heterostructures	Xe arc lamp	0.011 min^{-1}	RhB	Xing, Y., Que, W., Yin, X., Liu, X., Javed, H. A., Yang, Y., & Kong, L. B. (2015). Fabrication of Bi 2 Sn 2 O 7-ZnO heterostructures with enhanced photocatalytic activity. *RSC Advances*, *5*(35), 27576-27583
81	Bi_2 S_3/ZnO heterostructure	Visible-light	0.004 min^{-1}	MO	Bera, S., Ghosh, S., & Basu, R. N. (2018). Fabrication of Bi 2 S 3/ZnO heterostructures: an excellent photocatalyst for visible-light-driven hydrogen generation and photoelectrochemical properties. *New Journal of Chemistry*, *42*(1), 541-554.
82	(FGS)/ZnO nanocomposites	Visible-light	0.012 min^{-1}	safranin-T	Nenavathu, B. P., Kandula, S., & Verma, S. (2018). Visible-light-driven photocatalytic degradation of safranin-T dye using functionalized graphene oxide nanosheet (FGS)/ZnO nanocomposites. RSC advances, 8(35), 19659-19667.
83	ZnO/CdS hierarchical heterostructure	solar irradiation	0.0108 min^{-1}	RhB	Fang Xu, ab Yafei Yuan, a Huijuan Han,c Dapeng Wu, a Zhiyong Gaoab and Kai Jiang*ab CrystEngComm, 2012, 14, 3615–3622 Synthesis of ZnO/CdS hierarchical heterostructure with enhanced photocatalytic efficiency under nature sunlight CrystEngComm, 2012, 14, 3615–3622
84	$ZnGa_2O_4$ nanorod arrays	UV	0.0397 min^{-1}	RhB	Li, Z., Li, B., Liu, Z., Li, D., Ge, C., & Fang, Y. (2014). Controlled synthesis of $ZnGa_2O_4$ nanorod arrays from hexagonal ZnO microdishes and their photocatalytic activity on the degradation of RhB. RSC Advances, 4(89), 48590-48595
85	necklace-like ZnO–$ZnWO_4$	Hg lamp	0.035 min^{-1}	RhB	Hao, Y., Zhang, L., Zhang, Y., Zhao, L., & Zhang, B. (2017). Synthesis of pearl necklace-like ZnO–ZnWO 4 heterojunctions with enhanced photocatalytic degradation of Rhodamine B. RSC Advances, 7(42), 26179-26184

Sl. No.	Nanocomposite	UV/ Visible light activated	Rate constant	Used Dye	Reference
86	$ZnO/ZnAl_2O_4$ multi co-centric nanotubes	UV	0.0253 min^{-1}	MO	Nasr, M., Viter, R., Eid, C., Warmont, F., Habchi, R., Miele, P., & Bechelany, M. (2016). Synthesis of novel ZnO/ZnAl 2 O 4 multi co-centric nanotubes and their long-term stability in photocatalytic application. *RSC Advances*, *6*(105), 103692-103699.
87	ZnO/Cu2SnS3 nanorod	UV-light	0.0387 min^{-1}	RhB	Guo, Y., Yin, X., Yang, Y., & Que, W. (2016). Construction of ZnO/Cu 2 SnS 3 nanorod array films for enhanced photoelectrochemical and photocatalytic activity. *RSC Advances*, *6*(106), 104041-104048.

REFERENCES

Ahn C. H., Kim Y. Y., Kim D. C., Mohanta S. K., and Cho H. K. (2009). A comparative analysis of deep level emission in ZnO layers deposited by various methods. *Journal of Applied Physics* 105, 013502.

Alosfur et al. (2015). Modified microwave method for the synthesis of visible light-responsive TiO2/MWCNTs nanocatalysts. *Nanoscale Res. Lett.* 10, 346.

Ansari S. A., Khan M. A., Kalathil, S. Nisar A., Lee J., and Cho M. H. (2013). Oxygen vacancy induced band gap narrowing of ZnO nanostructures by an electrochemically active biofilm. *Nanoscale* 5, 9238–9246.

Banerjee S., Bhattacharyya P., Ghosh C. K. (2017). Charge carrier–LO phonon interaction in ZnO nanostructures: effect on photocatalytic activity and infrared optical constants. *Appl. Phys.* A.123. 640.

Bano N., Hussain I., Sawaf S., Alshammari A., and Saleemi F. (2017). Enhancement of external quantum efficiency and quality of heterojunction white LEDs by varying the size of ZnO nanorods. *Nanotechnology* 28. 245203.

Barbagiovanni E. G., Strano. V., Franzò G., Reitano R., Dahiya A. S., Poulin-Vittrant G., Alquier D., and Mirabella S. (2016). Universal model for defect-related visible luminescence in ZnO nanorods. *RSC Adv.* 6, 73170.

Bekeny C., Voss T., Gafsi H., and Gutowski J., Postels B., Kreye M., and Waag A. (2006). Origin of the near-band-edge photoluminescence emission in aqueous chemically grown ZnO nanorods. *Journal of Applied Physics* 100, 104317.

Bera A. and Basak D. (2011). Pd-nanoparticle-decorated ZnO nanowires: ultraviolet photosensitivity and photoluminescence properties. *Nanotechnology,* 22, 265501.

Bhattacharyya P.,·Bhattacharjee S., Bar M., Ghorai U. K., Pal M., Baitalik S., Ghosh C. K. (2018). Hedgehog ZnO/Ag heterostructure: an environment-friendly rare earth free potential material for cold-white light emission with high quantum yield. *Applied Physics A.* 124.782.

Birman J. L. (1959). Polarization of Fluorescence in CdS and ZnS single crystals, *Physical Review Letters*, 2, 157.

Brahma S., Khatei J., Sunkara S., Lo K-Y and Shivashankar S A. (2015). Self-assembled ZnO nanoparticles on ZnO microsheet: ultrafast synthesis and tunable photoluminescence properties. *J. Phys. D: Appl. Phys.* 48. 225305.

Camarda, P. et al. (2016). Luminescence mechanisms of defective ZnO nanoparticles. *Phys. Chem. Chem. Phys.,* 2016, 18, 16237—16244.

Chandrasekaran S., Chung J. S., Kim E. J., Hur S. H. (2016). Exploring complex structural evolution of graphene oxide/ZnO triangles and its impact on photoelectrochemical water splitting. *Chem. Eng. J.* 290.465–476.

Chen R. and McKeever S. W. S. (1997). *Theory of Thermoluminescence and Related Phenomenon.* World Scientific. Singapore.

Chen T, Xing G Z, Zhang Z, Chen H Y and Wu T. (2008). Tailoring the photoluminescence of ZnO nanowires using Au nanoparticles. *Nanotechnology.* 19. 435711.

Cheng C. W., Sie E. J., Liu B., Huan C. H. A., Sum T. C., Sun H. D., and Fan H. J. (2010). Surface plasmon enhanced band edge luminescence of ZnO nanorods by capping Au nanoparticles. *APL,* 96, 071107.

Connolly T. F. (1968) Semiconductors: Preparation, Crystal Growth, and Selected Properties. *Physica,* 39, 123 – 132.

Das S. Ghorai U. K., Dey R., Ghosh C. K. and Pal M. (2017). Novel multiple phosphorescence in nanostructured zinc oxide and calculations of correlated colour temperature. *Phys. Chem. Chem. Phys.,* 2017, 19, 22995—23006.

Dev A., Richters J. P., Sartor J., Kalt H., Gutowski J., and Voss T. (2011). Enhancement of the near-band-edge photoluminescence of ZnO nanowires: Important role of hydrogen incorporation versus plasmon resonances. *Applied Phys. Lett.* 98, 131111.

Dijken A. V., Meulenkamp E., Vanmaekelbergh D., Meijerink A. (2000). The Kinetics of the Radiative and Nonradiative Processes in Nanocrystalline ZnO Particles upon Photoexcitation. *J. Phys. Chem.* B, 104, 1715.

Djurišić A. B. et al. (2007). Defect emissions in ZnO nanostructures. *Nanotechnology* 18 095702

Dodd A., McKinley A., Tsuzuki T., Saunders M. (2009). Tailoring the photocatalytic activity of nanoparticulate zinc oxide by transition metal oxide doping. *Mater. Chem. Phys.* 114. 382–386.

Flores N. M., Pal U., Galeazzi R., and Sandoval A. (2014). Effects of morphology, surface area, and defect content on the photocatalytic dye degradation performance of ZnO nanostructures. *RSC Adv.* 4, 41099 – 41110.

Haynes J. R. (1960). Experimental proof of the existence of a new electronic complex in silicon. *Physics Rev. Lett.* 4, 361.

Ho C. H., Chen Y. J., and Li J. S. (2008). Optical properties of near band-edge transitions in well-aligned and tilted ZnO nanostructures. *J. Phys. D: Appl. Phys.* 41, 165410.

Hsu J. W. P., Tallant D. R., Simpson R. L., Missert N. A, and Copeland R. G. (2006). Luminescent properties of solution-grown ZnO nanorods. *Appl. Phys. Lett.* 88, 252103.

Jacopin G., Rigutti L., Bugallo A. D. L., Julien F. H., Baratto C., Comini E., Ferroni M. and Tchernycheva M. (2011). High degree of polarization of the near-bandedge photoluminescence in ZnO nanowires. *Nanoscale Research Letters,* 6, 501.

Janotti A. and Walle C. G. Van de. (2007). Native point defects in ZnO. *PRB,* 76, 165202.

Ji J., Boatner L. A., and Selim F. A. (2014). Donor characterization in ZnO by thermally stimulated luminescence. *Appl. Phys. Lett.* 105, 041102.

Jian X. et al. (2015). Enhancement in photoluminescence performance of carbon-decorated T-ZnO. *Nanotechnology.* 26. 125705.

John T. T., Priolkar K. R., Bessiere A., Sarode P. R., and Viana B. (2011). Effect of [OH–] Linkages on Luminescent Properties of ZnO Nanoparticles. *J. Phys. Chem.* C.115, 18070–18075).

Kavitha M. K., Gopinath and John H. (2015). Reduced graphene oxide–ZnO self-assembled films: tailoring the visible light photoconductivity by the intrinsic defect states in ZnO. *Phys. Chem. Chem. Phys.* 17. 14647–14655.

Khomyak V. V. et al. (2013). Annealing effect on the near-band edge emission of ZnO. *J. Phys. and Chem. of Solids.* 74. 291–297.

Liu B., Zhao X., Terashima C., Fujishima A., and Nakata K. (2014). Thermodynamic and kinetic analysis of heterogeneous photocatalysis for semiconductor systems. *PCCP,* 16, 8751 – 8760.

Liu G., Wang L., Yang H. G., Cheng H-M., and Lu G. Q. (M). (2010). Titania-based photocatalysts — crystal growth, doping and heterostructuring. *J. Mater. Chem.* 20, 831–843.

Look D. C., Hemsky J. W., and Sizelove J. R. (1999). *Residual Native Shallow Donor in ZnO PRL,* 82, 2552.

Mahmood M. A., Baruah S., Dutta J. (2011). Enhanced visible light photo catalysis by manganese doping or rapid crystallization with ZnO nanoparticles. *Mater. Chem. Phys.* 130, 531 – 535.

Mang A., Reimann K., Rübenacke St. (1995). Band gaps, crystal-field splitting, spin-orbit coupling, and exciton binding energies in ZnO under hydrostatic pressure. *Solid State Communications*, 94, 251.

Manzano C. V., Alegre D., Caballero-Calero O., Alén B., and Martín-González M. S. (2011). Synthesis and luminescence properties of electrodeposited ZnO films. *Journal of Applied Physics* 110, 043538.

Oulton R. F., Sorger V. J., Zentgraf T., Ma Ren-Min, Gladden C., Dai L., Bartal G. & Zhang X. (2009). Plasmon lasers at deep subwavelength scale. *Nature,* 461, 629.

Pal U., Meléndrez M. R. Chernov, V., and Barboza-Flores M. (2006). Thermoluminescence properties of ZnO and ZnO:Yb nanophosphors. *Applied Physica Letters* 89. 183118.

Pan N., Wang X., Li M., Li F., and Hou J. G. (2007). Strong Surface Effect on Cathodoluminescence of an Individual Tapered ZnO Nanorod. *J. Phys. Chem.* C. 111. 17265-17267.

Park C., Lee J., Chang W. S. (2015). Geometrical Separation of Defect States in ZnO Nanorods and Their Morphology-Dependent Correlation between Photoluminescence and Photoconductivity. *J. Phys. Chem.* C. 119. 16984−16990.

Phillips M. R. Gelhausen O., and Goldys E. M. (2004). Cathodoluminescence properties of zinc oxide nanoparticles. *Phys. stat. sol.* (a) 201, No. 2, 229–234.

Qi K. Cheng B., Yu J., Ho W. (2017). Review on the improvement of the photocatalytic and antibacterial activities of ZnO. *Journal of Alloys and Compounds* 727, 792 – 820.

Raji R., and Gopchandran K. G. (2017). ZnO:Cu nanorods with visible luminescence: Copper induced defect levels and its luminescence dynamics. *Mater. Res. Exp.,* 4, 025002.

Rauwel E. et al. (2011). Precursor-Dependent Blue-Green Photoluminescence Emission of ZnO Nanoparticles. *J. Phys. Chem.* C 2011, 115, 25227–25233.

Reynolds D. C., Look D. C., and Jogai B., Litton C. W., Collins T. C., Harsch W. and Cantwell G. (1998). Neutral-donor–bound-exciton complexes in ZnO crystals. *Physical Review* B, 57, 19.

Rodnyi P. A. and Khodyuk I. V. (2011). Optical and luminescence properties of zinc oxide. *Optics and Spectroscopy,* 111, 5, 776–785.

Rong M. Z., Zhang M. Q., Zheng Y. X., Zen H. M., Walter R., Friedrich K. (2001). Structure–property relationships of irradiation grafted nano-inorganic particle filled polypropylene composites. *Polymer.* 42. 167 – 83.

Rout C. S. and Rao C. N. R. (2008). Electroluminescence and rectifying properties of heterojunction LEDs based on ZnO nanorods. *Nanotechnology* 19 285203.

Samanta A., Gangopadhyay R., Ghosh C. K., Ray M. (2017). Enhanced photoluminescence from gold nanoparticle decorated polyaniline nanowire bundles. *RSC Advances* 7 (44), 27473-27479.

Singh J., Kumar P., Hui K. S., Hui K. N., Ramam K., Tiwari R. S., and Srivastava O. N. (2012). Synthesis, band-gap tuning, structural and optical investigations of Mg doped ZnO nanowires. *Cryst. Eng. Comm.* 14, 5898–5904.

Strzhemechny et al. (2003). Remote hydrogen plasma processing of ZnO single crystal surfaces. *J. Appl. Phys.*, Vol. 94, No. 7, 1 October 2003

Sun H., Zhang Q-F., and Wu J-L. (2006). Electroluminescence from ZnO nanorods with an n-ZnO/p-Si heterojunction structure. *Nanotechnology.* 17. 2271–2274.

Swiegers G. F. et al. (2012). Towards Hydrogen Energy: Progress on Catalysts for Water Splitting. *Aust. J. Chem.* 65. 577–582.

Wang J., Wang Z., Huang B., Ma Y., Liu Y., Zhang X., Dai Y. (2012). Oxygen Vacancy Induced Band-Gap Narrowing and Enhanced Visible Light Photocatalytic Activity of ZnO. *ACS Appl. Mater. Interfaces.* 4. 4024–4030.

Wei X. Q., Man B. Y., Liu M., Xue C. S., Zhuang H. Z., Yang C. (2007). Blue luminescent centers and microstructural evaluation by XPS and Raman in ZnO thin films annealed in vacuum, *N2 and O2 Physica* B 388. 145–152.

Wu K., Lu y., He H., Huang J., Zhao B., and Ye Z. (2011). Enhanced near band edge emission of ZnO via surface plasmon resonance of aluminum nanoparticles. *Journal of Applied Physics,* 110, 023510.

Xiao X. H., Ren F., Zhou X. D., Peng T. C., Wu W., Peng X. N, Yu X. F., and Jiang C. Z. (2010). Surface plasmon-enhanced light emission using silver nanoparticles embedded in ZnO. *Applied Physics Letter,* 97. 071909.

Xu C., Chun J., and Kim D. E. (2007). Electrical properties and near band edge emission of Bi-doped ZnO nanowires. *APL,* 90, 083113.

Yao Y-C. et al. (2016). Enhancing UV-emissions through optical and electronic dual-function tuning of Ag nanoparticles hybridized with n-ZnO nanorods/p-GaN heterojunction light-emitting diodes. *Nanoscale,* 8, 4463.

Yu W., Zhang J., Peng T. (2016). New insight into the enhanced photocatalytic activity of N-, C- and S-doped ZnO photocatalysts. *Appl. Catal.* B. 181.220–227.

Zhang Y. and Mu J. (2007). Controllable synthesis of flower- and rod-like ZnO nanostructures by simply tuning the ratio of sodium hydroxide to zinc acetate. *Nanotechnology,* 18, 75606.

Zhao J., Cu S., Zhan X. and Li X. (2018). Significantly enhanced UV luminescence by plasmonic metal on ZnO nanorods patterned by screen-printing. *Nanotechnology,* 29, 355703.

Zhou X., Kuang Q., Jiang Z-Y., Xie Z-X., Xu T., Huang R-B., Zheng L-S.,(2007). The Origin of Green Emission of ZnO Microcrystallites:

Surface-Dependent Light Emission Studied by Cathodoluminescence. *J. Phys. Chem.* C, 111, 12091-12093.

Zimmler M. A., Voss T., Ronning C. and Capasso F. (2009). Exciton-related lectroluminescence from ZnO nanowire light-emitting diodes. *Applied Physics Letters,* 94, 241120.

BIOGRAPHICAL SKETCH

Chandan Kumar Ghosh

Affiliation: School of Materials Science and Nanotechnology, Jadavpur University.

Education: PhD

Business Address: 188 Raja S. C. Mullick Road, Kolkata, West Bengal, India.

Research and Professional Experience: 15 years of research experience and 10 years teaching experience.

Professional Appointments: On 3rd December, 2009 by Jadavpur University

Honors: NET, GATE

Publications from the Last 3 Years:

1. Fast colorimetric detection of H_2O_2 by biogenic silver nanoparticles synthesized using Benincasa hispida fruit, Kaushik Roy, Chandan K. Sarkar, Chandan K. Ghosh, *Nanotechnology Reviews.* 5, 251 – 258 (2016).

2. Enhanced removal of dissolved aniline from water under combined system of nano zero-valent iron and Pseudomonas putida, Raj Shekhar Bose, Sayan Dey, Saswati Saha, Chandan Kr. Ghosh, Mahua Ghosh Chaudhuri, *Sustainable Water Resources Management.* 2, 143 – 159 (2016).
3. Novel green phosphorescence from pristine ZnO quantum dots: tuning of correlated color temperature, Sagnik Das, Chandan Kumar Ghosh, Rajib Dey, Mrinal Pal, *RSC Advances.* 6, 236 – 244 (2016).
4. NiO/Ag heterostructure: enhanced UV emission intensity, exchange interaction and photocatalytic activity, Santanab Majumder, Swarupananda Bhattacharjee, Chandan Kr. Ghosh, *RSC Advances.* 6, 56503 – 56510 (2016).
5. Fabrication of Magnetic Nanocrystals in Alcohol/Water mixed Solvents: Catalytic and Colloidal Property Evaluation, Srividhya J. Iyengar, Mathew Joy, A. Peer Mohamed, Swati Samanta, Chandan Kumar Ghosh, Swapankumar Ghosh, *RSC Advances* 6, 60845 – 60855 (2016).
6. Superb hydroxyl radical mediated biocidal effect induced antibacterial activity of tuned ZnO/chitosan type II heterostructure under dark, Soumik Podder, Suman Halder, Anirban Roychowdhury, Dipankar Das, Chandan Kumar Ghosh, *J. Nanoparticle Research* 18, 294 (2016).
7. Antibacterial mechanism of biogenic copper nanoparticles synthesized using Heliconia psittacorum leaf extract, K. Roy, C. K. Sarkar, C. K. Ghosh, *Nanotechnology Reviews* 5, 529 – 536 (2016).
8. Carbon doped ZnO thin film: unusual nonlinear variation in bandgap and electrical characteristic, Debabrata Sarkar, C. K. Ghosh and K. K. Chattopadhyay, *Applied Surface Science* 418, 252 – 257 (2017).
9. ROS mediated High Anti-Bacterial Efficacy of Strain Tolerant Layered Phase Pure Nano-Calcium Hydroxide, Aniruddah Samanta, Soumik Podder, Chandan Kumar Ghosh, Manjima Bhattacharya, Jiten Ghosh, Awadesh Kumar Mallik, Arjun Dey, Anoop Kumar

Mukhopadhyay, *Journal of the Mechanical Behavior of Biomedical Materials* 72, 110 – 128 (2017).

10. Enhanced Photoluminescence from Gold Nanoparticle Decorated Ployaniline Nanowire Bundles, Aniruddha Samanta, Rupali Gangopadhyay, Chandan Kumar Ghosh, Mallar Roy, *RSC Adv.* 7, 27473 – 27479 (2017).
11. Degradation of toxic textile dyes and detection of hazardous Hg^{2+} by low-cost bioengineered copper nanoparticles synthesized using Impatiens balsamina leaf extract, Kaushik Roy, Chandan K. Ghosh, Chandan K. Sarkar, *Materials Research Bulletin,* 94, 257 – 262 (2017).
12. Novel multiple phosphorescence in nanostructured Zinc oxide and calculation of correlated colour temperature, Sagnik Das, Uttam Kumar Ghorai, Rajib Dey, Chandan Kumar Ghosh, Mrinal Pal, *Physical Chemistry Chemical Physics,* 19, 22995 – 23006 (2017).
13. Charge carrier – LO phonon interaction in ZnO nanostructures: effect on photocatalytic activity and infrared optical constants, Shiny Banerjee, Puja Bhattacharyya and Chandan Kumar Ghosh, *Applied Physics A*, 123, 640 (2017).
14. Effect of human placental extract in the management of biofilm mediated drug resistance – A focus on wound management, Sutapa Goswami, Ratul Sarkar, Pritam Saha, Amit Maity, Tridib Sarkar, Debmalya Das, Piyali Dutta Chakraborty, Subhasri Bandopadhyay, Chandan Kumar Ghosh, Samnoy Karmakar, Tuhinadri Sen, *Microbial Pathogenesis,* 111, 307 – 315 (2017).
15. Multiferroicity around Verwey transition in Fe_3O_4 thin film, Shubhankar Mishra, Koushik Dey, Ujjal Chowdhury, Dipten Bhattacharya, Chandan Kumar Ghosh, Saurav Giri, *AIP Advances,* 7, 125015 (2017).
16. Non-inversion anisotropy energy in NiO coral structure: asymmetric hysteresis loop at room temperature, Swarupananda Bhattacharjee, Gopes Chandra Das, Anirban Roychowdhury, Dipankar Das, Chandan Kumar Ghosh, Dipten Bhattacharya, Pintu Sen, *Applied Surface Science,* 449, 389 – 398 (2018).

17. Facile synthesis of hierarchical nickel (III) oxide nanostructure: A synergistic remediating action towards water contaminants, Sayan Dey, Soumik Podder, A. Roychowdhury, Dipankar Das, Chandan Kumar Ghosh, *Journal of Environmental Management,* 211, 356 – 366 (2018).
18. Nonmonotonic particle-size-dependence of magnetoelectric coupling in strained nanosized particles of $BiFeO_3$, Sudipta Goswami, Dipten Bhattacharya, Chandan K. Ghosh, Barnali Ghosh, S. D. Kaushik, Vasudeva Sirguri and PSR Krishna, *Scientific Report,* 8, 3728 (2018).
19. Correlation between the dielectric and electrochemical properties of TiO_2-V_2O_5 nanocomposites for energy storage application, Apurba Ray, Atanu Roy, Swarupananda Bhattacharjee, Srikanta Jana, Chandan Kumar Ghosh, Chittranjan Sinha, Sachindranath Das, *Electrochimica Acta,* 266, 404 – 413 (2018).
20. Toxic Heavy Metal Ion Adsorption Kinetics of $Mg(OH)_2$ Nanostructures with Superb Efficacies, Dipak Chanda, Dipta Mukherjee, Pradip Das, Chandan Kumar Ghosh, Anoop Mukhopadhyay, *Materials Research Express,* 5, 075027 (2018).
21. Phase Pure, High Hardness, Biocompatable Calcium Silicates with Excellent Anti-bacterial and Biofilm Inhibition Efficacies for Endodontic and Orthopaedic Applications, Nilormi Biswas, Aniruddha Samanta, Soumik Podder, Chandan Kumar Ghosh, Jiten Ghosh, Mitun Das, Awadesh Mallik, Anoop Mukherjee, *Journal of the Mechanical Behavior of Biomedical Materials,* 86, 264 – 283 (2018).
22. Electron – phonon interaction to tune metal – semiconductor junction characteristic: ultralow potential barrier and less non-thermionic emission, Swarupananda Bhattacharjee, Arka Dey, Sayan Dey, Anirban Roychowdhury, Partha P. Ray, Dipankar Das, Gopes C. Das, Chandan K. Ghosh, *Physica B: Condensed Matter,* 547, 101 - 110 (2018).
23. The impact of morphology and concentration on crossover between antioxidant and pro-oxidant activity of MgO nanostructures,

Soumik Podder, Dipak Chanda, Anoop Kumar Mukhopadhyay, Arnab De, Bhaskar Das, Amalesh Samanta, John George Hardy, Chandan Kumar Ghosh, *Inorganic Chemistry,* 57 (20), 12727 – 12739 (2018).

24. Hedge-hog ZnO/Ag heterostructure: an environment friendly rare earth free potential material for cold white light emission with high quantum yield, Puja Bhattacharyya, Swarupananda Bhattacharjee, Manoranjan Bar, Uttam Kumar Ghorai, Mrinal Pal, Sujoy Baitalik and Chandan Kr. Ghosh, *Applied Physics A* 124, 782 (2018).
25. Zinc sulphide nanoparticle (nZnS): A novel nano-modulator for plant growth, Mala Thapa, Mukesh Singh, Chandan Kumar Ghosh, Prasanta Kumar Biswas, Abhishek Mukherjee, *Colloid and Interface Science Communications* 32 (2019) 100190.
26. Size-dependent antibacterial activity of copper nanoparticles againstXanthomonasoryzaepv.oryzae– A synthetic and mechanistic approach, Tapodhara Datta Majumdar, Mukesh Singh, Mala Thapa, Moumita Dutta, Abhishek Mukherjee, Chandan Kumar Ghosh, *Colloid and Interface Science Communications* 32 (2019) 100190.
27. Au nanoparticle-decorated aragonite microdumbbells for enhanced antibacterial and anticancer activities, Aniruddha Samanta, Soumik Podder, Murali Kumarasamy, Chandan Kumar Ghosh, Debrupa Lahiri, Partha Roy, Swarupananda Bhattacharjee, Jiten Ghosh, Anoop Kumar Mukhopadhyay, *Materials Science & Engineering* C 103 (2019) 109734.

Book Chapters:

1. Quantum Effect on Properties of Nanomaterials, CK Ghosh, *Introduction to Nano: Basic to Nanoscience and Nanotechnology*, page: 73-111, 2015, ISSN 1868 - 1212.
2. Advanced Characterization Techniques, CK Ghosh, *Introduction to Nano: Basic to Nanoscience and Nanotechnology*, page: 113-144, 2015, ISSN 1868 - 1212.

3. Growth Techniques and Characterization tools of Nanomaterials, CK Ghosh, Arka Dutta, *Nanotechnology: Synthesis to Application,* ISBN 978 – 1 – 1380 – 3273 – 6 CRC press (2017).
4. Introductory Quantum Mechanics for Nanoscience, CK Ghosh, *Nanotechnology: Synthesis to Application*, ISBN 978 – 1 – 1380 – 3273 – 6 CRC press (2017).
5. Environmental and Biological Applications of Nanomaterials, K Roy, CK Ghosh, *Nanotechnology: Synthesis to Application*, ISBN 978 – 1 – 1380 – 3273 – 6 CRC press (2017).
6. Biological Synthesis of Metallic Nanoparticles: A Green Alternative, K Roy, CK Ghosh, *Nanotechnology: Synthesis to Application*, ISBN 978 – 1 – 1380 – 3273 – 6, CRC press (2017).
7. Synthesis of Noble Metals: Chemical and Physical Routes, CK Ghosh, *Nanotechnology: Synthesis to Application*, ISBN 978 – 1 – 1380 – 3273 – 6 CRC press (2017).
8. Role of Nanomaterials in Food Preservation, Chandan Kumar Ghosh, Debabrata Bera, and Lakshmishri Roy, R. Prasad (ed.), *Microbial Nanobionics, Nanotechnology in the Life Sciences*, Springer Nature Switzerland AG 2019. https://doi.org/10.1007/978-3-030-16534-5_10

Editor of the Book: *Nanotechnology: Synthesis to Application*, ISBN 978 – 1 – 1380 – 3273 – 6 CRC press (2017).

INDEX

F

G

H

I

J

K

L

M

N

O

P

Q

R

S

T

U

V

W

X

Y

Z

Related Nova Publications

PHOTOCATALYSIS: PERSPECTIVE, MECHANISM, AND APPLICATIONS

EDITORS: Preeti Singh, M. M. Abdullah, MD, Mudasir Ahmad, and Saiqa Ikram

SERIES: Nanotechnology Science and Technology

BOOK DESCRIPTION: This book covers all important topics of photocatalysis in simple language with clear presentation. Nanostructures are continuously improving the functional characteristics of the material because of enhanced surface to volume ratio.

HARDCOVER ISBN: 978-1-53616-044-4
RETAIL PRICE: $230

INTRODUCTION OF FORENSIC NANOTECHNOLOGY AS FUTURE ARMOUR

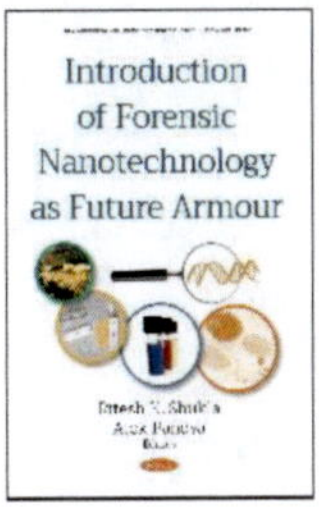

EDITORS: Ritesh Kumar Shukla and Alok Pandya

SERIES: Nanotechnology Science and Technology

BOOK DESCRIPTION: Introduction of Forensic Nanotechnology as Future Armour is the first comprehensive book to consider both fundamental and applied aspects of forensic nanotechnology.

HARDCOVER ISBN: 978-1-53616-040-6
RETAIL PRICE: $195

To see a complete list of Nova publications, please visit our website at www.novapublishers.com

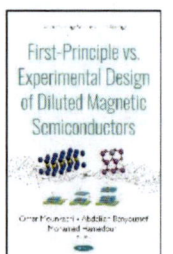